Atomic Layer Deposition

Atomic Layer Deposition

Editor

David C. Cameron

MDPI • Basel • Beijing • Wuhan • Barcelona • Belgrade • Manchester • Tokyo • Cluj • Tianjin

Editor
David C. Cameron
Department of Physical Electronics,
Masaryk University
Czech Republic

Editorial Office
MDPI
St. Alban-Anlage 66
4052 Basel, Switzerland

This is a reprint of articles from the Special Issue published online in the open access journal *Coatings* (ISSN 2079-6412) (available at: https://www.mdpi.com/journal/coatings/special_issues/atomic_depos).

For citation purposes, cite each article independently as indicated on the article page online and as indicated below:

LastName, A.A.; LastName, B.B.; LastName, C.C. Article Title. *Journal Name* **Year**, *Article Number*, Page Range.

ISBN 978-3-03936-652-1 (Pbk)
ISBN 978-3-03936-653-8 (PDF)

© 2020 by the authors. Articles in this book are Open Access and distributed under the Creative Commons Attribution (CC BY) license, which allows users to download, copy and build upon published articles, as long as the author and publisher are properly credited, which ensures maximum dissemination and a wider impact of our publications.

The book as a whole is distributed by MDPI under the terms and conditions of the Creative Commons license CC BY-NC-ND.

Contents

About the Editor vii

Preface to "Atomic Layer Deposition" ix

César Masse de la Huerta, Viet Huong Nguyen, Jean-Marc Dedulle, Daniel Bellet, Carmen Jiménez and David Muñoz-Rojas
Influence of the Geometric Parameters on the Deposition Mode in Spatial Atomic Layer Deposition: A Novel Approach to Area-Selective Deposition
Reprinted from: *Coatings* **2019**, *9*, 5, doi:10.3390/coatings9010005 1

Wen-Jen Lee and Yong-Han Chang
Growth without Postannealing of Monoclinic VO_2 Thin Film by Atomic Layer Deposition Using VCl_4 as Precursor
Reprinted from: *Coatings* **2018**, *8*, 431, doi:10.3390/coatings8120431 15

Robert Müller, Lilit Ghazaryan, Paul Schenk, Sabrina Wolleb, Vivek Beladiya, Felix Otto, Norbert Kaiser, Andreas Tünnermann, Torsten Fritz and Adriana Szeghalmi
Growth of Atomic Layer Deposited Ruthenium and Its Optical Properties at Short Wavelengths Using $Ru(EtCp)_2$ and Oxygen
Reprinted from: *Coatings* **2018**, *8*, 413, doi:10.3390/coatings8110413 27

Richard Krumpolec, Tomáš Homola, David C. Cameron, Josef Humlíček, Ondřej Caha, Karla Kuldová, Raul Zazpe, Jan Přikryl and Jan M. Macak
Structural and Optical Properties of Luminescent Copper(I) Chloride Thin Films Deposited by Sequentially Pulsed Chemical Vapour Deposition
Reprinted from: *Coatings* **2018**, *8*, 369, doi:10.3390/coatings8100369 41

Tatiana V. Ivanova, Tomáš Homola, Anton Bryukvin and David C. Cameron
Catalytic Performance of Ag_2O and Ag Doped CeO_2 Prepared by Atomic Layer Deposition for Diesel Soot Oxidation
Reprinted from: *Coatings* **2018**, *8*, 237, doi:10.3390/coatings8070237 57

Luis Javier Fernández-Menéndez, Ana Silvia González, Víctor Vega and Víctor Manuel de la Prida
Electrostatic Supercapacitors by Atomic Layer Deposition on Nanoporous Anodic Alumina Templates for Environmentally Sustainable Energy Storage
Reprinted from: *Coatings* **2018**, *8*, 403, doi:10.3390/coatings8110403 73

Miia Mäntymäki, Mikko Ritala and Markku Leskelä
Metal Fluorides as Lithium-Ion Battery Materials: An Atomic Layer Deposition Perspective
Reprinted from: *Coatings* **2018**, *8*, 277, doi:10.3390/coatings8080277 91

About the Editor

David Cameron received his PhD from the University of Glasgow (UK) in the field of molecular beam epitaxy. He joined the Royal Signals and Radar Establishment (UK) in 1979 and moved to Dublin City University, School of Electronic Engineering, (Ireland) in 1982 where he set up the Thin Film Materials Research Laboratory and became an Associate Professor. He moved to Lappeenranta University of Technology (Finland) in 2004 as Professor of Material Technology and set up and led the Advanced Surface Technology Research Laboratory (ASTRaL). He began work with the Department of Physical Electronics at Masaryk University (MU), Brno (Czech Republic) in 2017 as a Research Scientist to develop expertise in atomic layer deposition. He is the author of 143 peer-reviewed journal papers and 1 book (to date), with an h factor of 36 (Google Scholar). His research career has focused on thin film deposition—plasma CVD, magnetron sputtering, sol-gel deposition, and atomic layer deposition. He retired from MU in 2020.

Preface to "Atomic Layer Deposition"

Atomic layer deposition (ALD) is a process that is renowned for its ability to produce films with unrivaled thickness control, conformability to three-dimensional structures, control over composition, and versatility in the range of materials it can produce from quaternary compounds to elemental metals. It has expanded from a small-scale batch process to large scale production now also including continuous processing—spatial ALD. It has matured into an industrial technology essential for many areas of materials science and engineering from microelectronics to corrosion protection. Its attributes make it a key technology in studying new materials and structures over an enormous range of applications. This Special Issue contains six research articles and one review article that illustrate the breadth of these applications from energy storage in batteries or supercapacitors to catalysis via x-ray, UV, and visible optics. They deal with the details of the ALD processes that produce these various films, the properties of the devices, and simulations that illustrate how the ALD system configuration affects the deposition process. In their research article, de la Huerta et al. explored gas flow issues in atmospheric pressure spatial ALD by computational fluid dynamics. They demonstrate the influence of the size and uniformity of the spacing between the coating head and the substrate and how it affects the transition from ALD to CVD behavior. The importance of exhaust efficiency in removing the reaction byproducts in the transition from ALD to CVD has also been explored. They have shown how control of the substrate-head spacing could be used as a method to obtain selective area deposition. Lee and Chang used the inorganic precursor VCl_4 to directly deposit crystalline VO_2. The thermal stability of this inorganic precursor for V compared to the typically used metal-organic precursors allows a high enough deposition temperature to render the usual post-annealing process unnecessary. The films show a transition from monoclinic to rutile crystal structure between 30 °C and 90 °C corresponding to a semiconductor to metal phase transition. Müller et al. deposited Ru metal films for XUV and x-ray optical applications and compared the results with films produced by magnetron sputtering. The ALD films show lower stress but somewhat higher roughness. This, together with the existence of a thin oxide surface film, leads to lower reflectance in the wavelength range of interest. The process needs further development to improve the optical properties to fully exploit the advantages of ALD, such as 3D conformality, in short-wavelength optics. Krumpolec et al. investigated the deposition of the wide bandgap semiconductor γ-CuCl using pyridine hydrochloride and a Cu metal-organic compound as precursors. CuCl is of interest for UV optical applications because of its bandgap and high exciton binding energy. They have shown that films of CuCl could be deposited without any Cu^{2+} content and could be protected against hydrolysis by atmospheric moisture using an Al_2O_3 capping layer. Ivanova et al. investigated the use of silver oxide and silver-doped CeO_2 ALD films for catalytic oxidation of diesel exhaust soot. In their work, a 1:10 composition of Ag in CeO_2 deposited on stainless steel had the best performance, with complete combustion of the soot at 390 °C, lower than for pure Ag_2O. The films showed consistent performance in repetitive tests whereas Ag_2O showed fast deterioration. The activity was found to be caused by oxygen species bound to Ag^+ sites. Fernández-Menéndez et al. explored a new manufacturing sequence for supercapacitors based on porous alumina with Al_2O_3:Zn conducting contact material and a dielectric layer consisting of either Al_2O_3 or a $SiO_2/TiO_2/SiO_2$ triple layer. They show that the device containing the Al_2O_3 dielectric is better in terms of lower internal resistance, lower leakage current, and higher breakdown voltage. Nevertheless, the overall performance still needs improvement with lower resistance internal contact layers and better external contacts particularly required. The films showed characteristic free and bound excitonic emissions and structure in photo-luminescence and optical reflectance, respectively. The review by Mäntymäki et al. covered the basics of Li-ion batteries and a discussion of metal fluorides as Li-ion battery materials, used as electrodes, electrode-electrolyte interphase layers, and solid electrolytes. The review demonstrates that metal fluorides have interesting properties that could provide high voltage and high capacity alternatives to oxide-based materials and that these are worthy of further research. They propose that the advantages of ALD processes, namely conformal coatings with precise thickness control for ultra-thin films could answer the demands of battery materials. They review the previous work on ALD of metal fluorides and suggest that these are ripe for future investigations. This paper will prove valuable to those investigating metal fluorides both for battery and other applications.

David C. Cameron
Editor

Article

Influence of the Geometric Parameters on the Deposition Mode in Spatial Atomic Layer Deposition: A Novel Approach to Area-Selective Deposition

César Masse de la Huerta, Viet Huong Nguyen, Jean-Marc Dedulle, Daniel Bellet, Carmen Jiménez and David Muñoz-Rojas *

Université Grenoble Alpes, CNRS, Grenoble INP, LMGP, 38000 Grenoble, France;
cesar.masse@grenoble-inp.fr (C.M.d.l.H.); viet-huong.nguyen@grenoble-inp.fr (V.H.N.);
jean-marc.dedulle@grenoble-inp.fr (J.-M.D.); daniel.bellet@grenoble-inp.fr (D.B.);
carmen.jimenez@grenoble-inp.fr (C.J.)
* Correspondence: david.munoz-rojas@grenoble-inp.fr; Tel.: +33-456-529-337

Received: 3 November 2018; Accepted: 19 December 2018; Published: 22 December 2018

Abstract: Within the materials deposition techniques, Spatial Atomic Layer Deposition (SALD) is gaining momentum since it is a high throughput and low-cost alternative to conventional atomic layer deposition (ALD). SALD relies on a physical separation (rather than temporal separation, as is the case in conventional ALD) of gas-diluted reactants over the surface of the substrate by a region containing an inert gas. Thus, fluid dynamics play a role in SALD since precursor intermixing must be avoided in order to have surface-limited reactions leading to ALD growth, as opposed to chemical vapor deposition growth (CVD). Fluid dynamics in SALD mainly depends on the geometry of the reactor and its components. To quantify and understand the parameters that may influence the deposition of films in SALD, the present contribution describes a Computational Fluid Dynamics simulation that was coupled, using Comsol Multiphysics®, with concentration diffusion and temperature-based surface chemical reactions to evaluate how different parameters influence precursor spatial separation. In particular, we have used the simulation of a close-proximity SALD reactor based on an injector manifold head. We show the effect of certain parameters in our system on the efficiency of the gas separation. Our results show that the injector head-substrate distance (also called deposition gap) needs to be carefully adjusted to prevent precursor intermixing and thus CVD growth. We also demonstrate that hindered flow due to a non-efficient evacuation of the flows through the head leads to precursor intermixing. Finally, we show that precursor intermixing can be used to perform area-selective deposition.

Keywords: spatial atomic layer deposition (SALD); computational fluid dynamics; surface reaction; thin films; ALD deposition; CVD deposition; area-selective deposition

1. Introduction

Atomic layer deposition (ALD) is a material deposition process that allows for a homogeneous, conformal thin film deposition with a nanometric thickness control. ALD is a type of chemical vapor deposition (CVD) method characterized by self-limited surface reactions. In ALD, instead of allowing a simultaneous presence of the reactants as is the case in the conventional CVD processes, a sequential exposure of the substrate to different reactants is needed to perform a chemical reaction with the substrate surface. A typical ALD cycle includes, periodically in time, exposure to a precursor, a purging step, an oxidant, and a second purging step. Vacuum processing is generally used in ALD in order to accelerate the purge steps and due to the traditional use of ALD in the microelectronics industries [1]. ALD cycles are characterized by having a defined growth per cycle (GPC) that depends on the chemical properties of the precursor, the temperature of the surface, and the reactor geometry. To attain a certain thickness, a determined number of cycles is performed. A review of the origins of ALD and a recommended reading list can be found in Reference [2].

Spatial Atomic Layer Deposition (SALD) is a technique based on the same principles of conventional (also called temporal) ALD, whose popularity is growing among the materials research community due to the fast deposition rates it offers, ranging from 20 to 40 nm/min, and to the large-area deposition capabilities at atmospheric pressure, and even in the open air, thus making it very appealing for the industry [3,4]. In addition, it offers the possibility of area-selective deposition [5,6], simplicity of installation, and allows depositing high-quality materials with a higher throughput than ALD.

In Spatial ALD, the main difference with respect to conventional ALD resides on a spatial separation of continuously injected reactants. Instead of defining each step by a time separation, and to achieve the same chemical half reactions that take place during the temporal ALD cycles, in SALD, precursors are injected continuously in different spatial regions of the reactor and the substrate is exposed alternately to the different flows, separating each subsequent exposure with an intermediate exposure to an inert gas, to purge the substrate of the half-reaction by-products, and/or excess of precursor. This spatially separated exposure of the substrate is equivalent to the temporal ALD cycles and achieves comparable materials properties when the materials deposited are not sensitive to the atmosphere [7]. SALD has been tested before by several groups to deposit a wide variety of functional oxides in a homogeneous and conformal manner, in many cases taking place at atmospheric pressure [8–10].

Numerous approaches have been explored to successfully generate the mentioned spatial regions needed, without intermixing the gaseous precursors in SALD [9–12]. Specifically, the approach used in our laboratory (a home-made system presented and explained in detail in [4]) is based on a patent published by Kodak [13] that led to the publication of scientific papers using the spatial separation ALD concept by the same group from 2008 [4]. The system relies on a deposition head with linear gas outlets that injects above the substrate surface a given flow and concentration of reactants within an inert carrier gas. The substrate is usually placed at a close distance (50–200 µm) during deposition, henceforth referred to as deposition gap. Such technique is commonly known as "close-proximity approach" since a small deposition gap value is necessary to prevent precursor intermixing across the inert gas region, thus avoiding a CVD regime deposition, i.e. reaction of the precursors reaching the substrate surface.

The spatial attribute of SALD gives many advantages with respect to temporal ALD, but due to the small value for the gap deposition needed, and to the fact that SALD is generally based on a mechanical displacement of the substrate, to fully exploit its advantages, a high mechanical and geometrical precision of the system needs to be carefully used. Furthermore, since our SALD approach does not rely on a chamber to be filled with the gases but rather on a continuous gas flow directed towards the surface of the substrate, the flow of such gases needs to be optimized as well to control the deposition conditions and to improve the homogeneity of the deposited film. Full control over these parameters is thus needed to enable a fast, large-area deposition with SALD.

Controlling the deposition gap can improve the versatility of the SALD, allowing it to tune the properties of the deposited film. In a CVD-like regime, films can be deposited in a fast way, but compactness, homogeneity, and control of thickness may be sacrificed. In an ALD regime, surface reactions on the substrate are self-limited, yielding a slower deposition rate, but a high conformality, homogeneity, and a good control of thickness are obtained in return. For the CVD regime to occur, intermixing of reactants must take place, yielding reactions above the surface before the precursors can reach it and be physi/chemisorbed. In contrast, in the ALD regime, the reactants must be chemisorbed, and ideally, saturate the surface before introducing the second reactant that leads to a complete surface reaction, thus creating a monolayer of the product. This key difference can be tuned arbitrarily in close-proximity SALD systems in which the deposition gap can be mechanically changed and thus it may provide versatility to tune the regime even in the middle of a deposition process [14]. A schematic of our injection head can be observed in Figure 1a, where the arrows represent the outlets and exhausts of the injected gases. The black arrows represent the inert carrier gas (I) that serves the purpose of confining the reactants and avoid intermixing. The white arrows represent exhausts (E) to which the gases can flow after being injected towards the surface of the substrate. The colored arrows represent the outlets of gases that contain the reactants used to create the surface reaction: the oxidant precursor (OP) in red and the ALD metal precursor (MP) in blue. Figure 1b shows the equivalent geometry used to perform the simulations. It is the region of interest from the original SALD schematic and corresponds to the region surrounded by a red dotted line in Figure 1a. The bottom-most line would represent the surface of the substrate and this line will be used as the place where surface reactions take place.

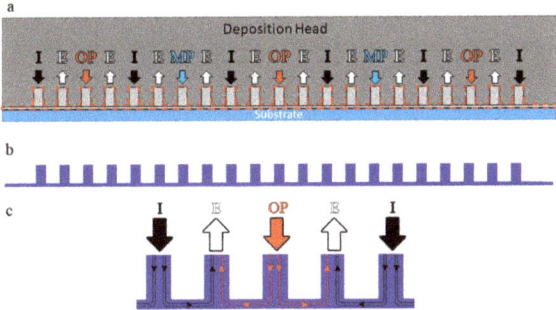

Figure 1. (a) Schematic of the cross-section of the deposition head in the spatial atomic layer deposition (SALD) system at the Laboratoire des Matériaux et du Génie Physique (LMGP). The gray section represents the deposition head on the system, while the blue section below represents a substrate. These two are separated by a space that corresponds to the deposition gap. (b) Equivalent geometry used for simulations used to compute all the phenomena in SALD regarding flows, concentrations, and reactions. (c) Close-up to the region of the OP, showing the expected flow lines and directions of the gaseous mixture of the SALD.

In this work, we have used computational fluid dynamics (CFD) simulations and coupled them with a surface reaction chemistry computation using Comsol Multiphysics® 5.3 in order to model our homemade close-proximity SALD deposition system. Accordingly, the influence of different parameters of our SALD system on the intermixing of the gaseous precursors has been studied. A quantification of CVD or ALD deposition regime, which can affect the quality and homogeneity of the film deposited, has been calculated. The gas flows in the setup are studied for a controlled separation of species in the reactor, and adjusted to control the appearance of a CVD component in the deposition, as has indeed been reported in the past [4,11]. We show that the capability of controlling the deposition regime can be indeed used to have area-selective deposition with a close-proximity SALD. Finally, tolerances on the geometry and on the mechanical design of the system are presented as a guide towards a correct mechanical design of a versatile and reproducible SALD deposition system.

2. Methods and Processes

To calculate the influence of the deposition gap on the growth regime in our system, a Comsol Multiphysics® simulation, which couples CFD with concentrated species diffusion and with surface chemical reactions, was used. For this, an equivalent geometry that includes the gap and the outlets and exhausts of the deposition head was used to compute the gas flow to then couple it with the reactant concentration distribution in the flow, and ultimately, with the surface chemical reaction that happens on the substrate surface. A schematic representation of this equivalent geometry is shown in Figure 1b. Reference [15] presents a numerical study of a flow in a close-proximity system and concludes that, in the deposition gaps usually used in close-proximity SALD (typically 50–200 µm), the Péclet number is low. This means that the transport of reactants is dominated by diffusion rather than by convection. We thus expect that the diffusion will play a key role in the behavior of our close-proximity system as well.

As said previously, this work simulates the gas flow that occurs with the gas outlets and the exhausts, as well as their influence on the distribution of reactant concentration on the substrate surface as a function of certain parameters of the system. This allows us to elucidate how such parameters influence on the existence of a CVD regime (i.e., precursor intermixing taking place). The simulation involved the study of the efficiency of the system to prevent regions where CVD could occur by calculating the conditions on which both reactants can be in contact prior to being adsorbed onto the substrate. The simulation has not considered a head/substrate relative movement during the deposition since the scope is only the efficiency of gas separation as a function of the different parameters studied.

Thus, the concentration of each reactant was computed at the immediate region above the surface of the substrate and the section in which both reactants are present was quantified (vide infra).

The SALD reactants consist of an ALD metal precursor and an oxidant. In the case of this work, Diethylzinc (DEZ) was used as the metal precursor, and water as the oxidant. To maintain the physical conditions as close as possible to the real conditions used, gas inputs of the injection head were taken as 300, 450, and 900 sccm for the precursor (DEZ), oxidant (H_2O), and nitrogen separation, respectively. These values are in accordance with what is commonly used when depositing zinc oxide (ZnO) films with our system [7]. The diffusion coefficient of the reactants used was in the order of magnitude of 10^{-3} $m^2 \cdot s^{-1}$ [16].

For the experimental validation of the simulations described here, ZnO was deposited using the SALD system at LMGP. A nitrogen flow carrying DEZ as precursor and water as oxidant to perform the surface reaction were used. A flow of pure nitrogen was used as the gas barrier between the reactants. The flows used were 300 sccm for DEZ, 450 sccm for water, and 900 sccm for the nitrogen separation. The substrate was heated during deposition at 200 °C. The scanning speed of the substrate was 50 mm/s.

In order to simulate the reactions that happen during a SALD deposition, it is important to understand the simulation workflow. The ultimate objective is to study the surface reaction of species onto the substrate surface. For this, we first perform a zero-dimensional simulation of the CVD reaction, following the equation:

$$DEZ + H_2O \xrightarrow{R_{AB}} CVD\ adsorbed\ film \quad (1)$$

which assumes that whenever DEZ and water molecules meet above the substrate they will react to form ZnO. Equation (1) allows quantifying the amount of deposition obtained in a CVD regime.

Then, a CFD analysis of the flows in the deposition gap was performed and the results are presented in the geometry shown in Figure 1c. This computation yields the velocity and pressure of the flow at every point. Next, the velocity component of the CFD results was used to calculate a diffusion of concentrated species along the geometry. As a result, the presence of each reactant at any point of the geometry can be obtained. Finally, a surface concentration due to the CVD regime reaction is calculated using Equation (1) and the pressure obtained in the CFD computation. For such a surface reaction to happen, both reactants (in this case DEZ and H_2O) need to be present at a given time. Thus, the final surface reaction will yield the amount of CVD regime deposition.

3. Results and Discussion

3.1. Evaluation of the Velocity Profile and Pressure in the Head-Substrate Gap

Figure 2a shows the velocity profile obtained for a geometry in which the deposition gap was fixed at 150 µm, a value that is commonly used in real depositions and using the gas flow values previously mentioned. The profile shows the expected flow from the gas outlets to the exhausts, but it shows a non-zero value on the lateral outflow regions of the simulation, which would indicate that the exhaust at the injection head may not be as efficient as required. The simulation shows a maximum velocity of ~2 m·s^{-1} at the narrowest point on the fluid's path from the inlet to the exhaust. The pressure profile is also shown in Figure 2b and it presents a gradient of pressure from the corner of the gas outlets to the corner of the exhausts. For comparison, Reference [17] presents a similar study on the deposition gap of another close-proximity system and concludes that a 2 mm gap with a "low pumping" at the exhausts would achieve a good spatial separation of the reactants. While their system also relies on inlets, exhausts and a deposition gap, the geometry is not planar, as opposed to our system. It is also important to notice that their system works at a pressure lower than the atmospheric pressure, which prevents confinement of the gaseous flows, and that the geometry of the system may enhance diffusion processes, and therefore, enhance intermixing when working at close-proximity. In our case, the pressure achieved by the geometry (especially in the corners close to the exhaust) is helpful to induce a flow towards the exhaust, hence improving the convective flow of the reactants to the exhaust and reducing the diffusive flow. This in turn contributes to preventing precursor intermixing.

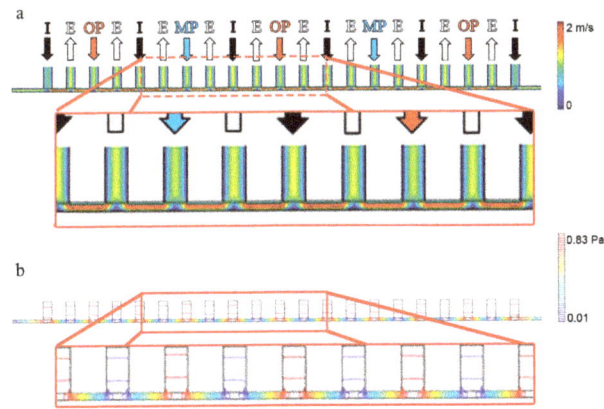

Figure 2. Computational Fluid Dynamics calculation made with Comsol Multiphysics® to represent the flow of the region of interest in the SALD geometry: (**a**) The velocity increases in the deposition gap, given the close proximity, and (**b**) the pressure increases under each of the outlets as it enters the deposition gap.

3.2. Study of the Effect of the Head-Substrate Deposition Gap on the Deposition Mode

Using the calculation of the velocity and the pressure obtained in the CFD computation, the concentration of each reactant was calculated along the gap, and more importantly, in the immediate region above the surface of the substrate for different deposition gaps. Figure 3 shows the concentration of the different reactants obtained along the length of the substrate (30 mm).

One can observe that, for a gap of 150 µm, the separation of reactants is well achieved (Figure 3a). In the concentration plot, the colors that represent the concentration of reactants are well separated along the geometry, with dark blue color (i.e., no precursor) below and next to the inert gas inlet channels, thus indicating a well-defined ALD regime. On the other hand, when the deposition gap is increased to 750 µm, the concentration is no longer well defined, as shown by the light blue color in the regions between each precursor. Thus, with this deposition gap, the deposition occurs in CVD regime (Figure 3b). In addition, to clearly compare different deposition gaps, first we plot the concentration of reactants in the carrier gases at the immediate region above the substrate, and then we quantified the amount of "overlap" between the plot of each precursor (indicated by the gray region in Figure 3). Again, a deposition gap of 150 µm shows almost no overlap, whereas a deposition gap of 750 µm clearly shows a much greater overlap.

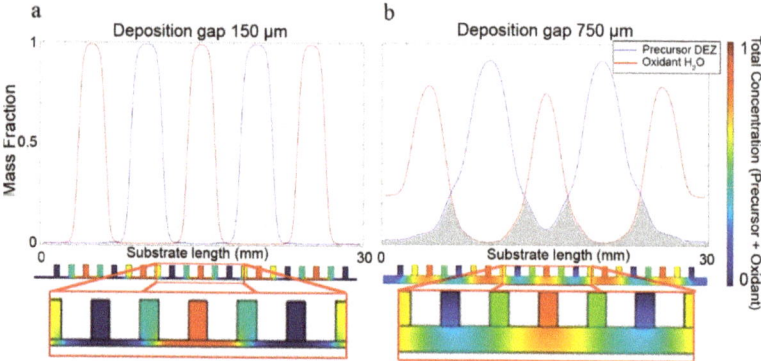

Figure 3. Concentration plot along the immediate region above the substrate for a deposition gap of (**a**) 150 and (**b**) 750 μm. Under each plot, a 2D plot along the whole geometry of the gap is shown, with a color code that corresponds to the concentration of reactants along the whole gap geometry.

Furthermore, in order to quantify the overlap for each deposition gap value, we calculated the ratio between the area under the curve of the overlap (where both reactants are present at the same time represented by the gray region in Figure 3) and the area under the curve of the sum of both concentrations, yielding a "percentage of overlap".

With this approach, further calculations of the percentage of overlap as a function of the deposition gap were made and are shown in Figure 4. This graph shows that, as the gap is increased, the overlap increases as well. Nevertheless, at a deposition gap value of around 750 μm the gases mainly flow out through the sides of the head, rather than being confined on top of the substrate, and therefore there is a change in the slope of the curve and a lower overlap than it would be expected takes place.

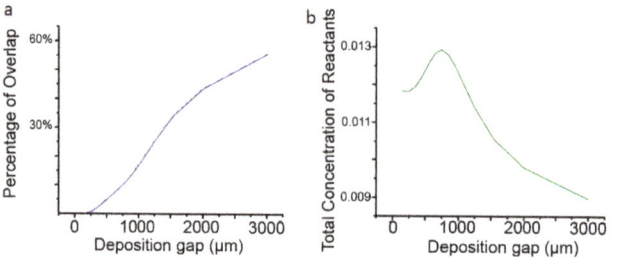

Figure 4. (**a**) The percentage of overlap, i.e., the percentage in which there exist both reactants at the same time, creating thus a CVD regime reaction on the surface of the substrate; (**b**) The total concentration of all the gases (both separated and overlapped) at the immediate area above the substrate vs. head-surface deposition gap.

Figure 4 also shows a plot of the total concentration of all reactants in the immediate region above the surface of the substrate, as measured by integrating the sum of the concentration of both reactants at the line of the substrate. The concentration reaches a maximum at ~750 μm, which means that, at that deposition gap, the flow is preferentially directed towards the surface of the substrate, rather than at the lateral outflow regions. Nevertheless, the overlap percentage at that deposition gap is ~13%, which indicates the existence of a CVD component taking place. Logically, as the deposition gap increases, the flow tends to be directed to the lateral outflow regions rather than be captured at the injection head exhaust, making the extraction of the surplus of reactants difficult, leading to a release of chemicals to the atmosphere, which should, of course, be avoided.

Study of the CVD Mode as a Consequence of Precursor Crosstalk

On a SALD reactor, reactions take place on the surface of the substrate, thus generating the desired films. Such reactions, in principle through chemisorption, take place as a consequence of the presence of a certain concentration of a reactant above the surface, and this depends on the pressure at each point, and of the substrate surface temperature.

In the case of an ALD deposition, the film deposition happens in two sequential half-reactions on the surface. Each half-reaction is self-limited to the surface sites available. In the case of SALD, if we assume a correct separation of the reactants, a static substrate, and a correct extraction of the reactant surplus, regardless of the time of injection of gases, the concentration of adsorbed reactant molecules in the surface should not be higher than the concentration of available sites. The surface coverage (θ) can be explained with the help of a Langmuir isotherm [18]:

$$\theta = \frac{k_{ads}P}{k_{des} + k_{ads}P} = \frac{\text{(number of occupied sites)}}{\text{(total number of sites)}}; \ 0 \leq \theta \leq 1 \quad (2)$$

where k_{ads} and k_{des} represent, respectively, the rate of adsorption/desorption of the reactant to/from the surface, and P is the precursor partial pressure.

Furthermore, the reaction probability depends on the surface coverage, since, as more sites are occupied, the sticking probability β of a reactant diminishes:

$$\beta = \beta_0(1 - \theta) \quad (3)$$

where β_0 is the "bare reaction probability", given by the intrinsic properties of the reactant. Equation (3) is taken from [19], which also mentions that the saturation time of the reactant (our ALD precursor in this case) is inversely proportional to the precursor pressure and to β_0.

In contrast to an ALD deposition, in a CVD deposition, the reactants are present in the gas simultaneously, which will induce both a surface chemical reaction, due to both chemisorption and thermal activation of the reaction, and at a lower rate, at the gas phase. In the substrate surface, this CVD reaction will induce a competition among the reactants for the available surface sites. The reactants may react to be chemisorbed or may be desorbed from the surface. To describe this phenomenon, the Langmuir-Hinshelwood reaction rate equation can be used [18]:

$$R_{AB} = \frac{k_{react}K_AK_BP_AP_B}{(1 + K_AP_A + K_BP_B)} \quad (4)$$

where k_{react}, K_A, and K_B are reaction constants corresponding to the whole reaction, and to reactant A and B, respectively, and P_A and P_B are the partial pressures of each reactant. To express the partial pressure of each reactant in terms of its concentration, we can assume an ideal gas behavior and express them as:

$$P_A = \frac{n_A}{n}P = x_AP \quad (5)$$

and

$$P_B = \frac{n_B}{n}P = x_BP \quad (6)$$

where n_A and n_B are the numbers of moles of each reactant, n is the total number of moles of the solution and x_A and x_B are the partial fractions of each reactant. Hence, substituting in Equation (4), we can obtain:

$$R_{AB} = \frac{k_{react}K_AK_Bx_Ax_BP}{\left(\frac{1}{P} + K_Ax_A + K_Bx_B\right)} \quad (7)$$

In Equation (7), the reaction rate of each reactant is considered, as well as a reaction rate of the reaction as a whole. This equation indicates that the reaction rate is not self-limited and will, therefore, continue as long as there is a non-zero concentration for both reactants. It also implies that if at any point the mass fraction of any of the reactants is zero, the reaction rate will also be zero and hence, no reaction would occur.

With this in mind, several time-dependent simulations were carried out in Comsol Multiphysics® to observe the appearance of a CVD deposition regime for different values of the deposition gap. The surface concentration obtained in such CVD regime is characterized by Equation (1) (Section 2), while the reaction rate R_{AB} comes from Equation (7). The pressure and the mass fraction of each reactant are calculated before the time-dependent surface chemistry reaction study, in the laminar flow study presented in Section 3.1, and in the concentration simulation of each reactant in the flow, respectively. As the CVD reaction is not self-limited, such time-dependent simulation was limited to 1 s. The reaction rate used for the CVD surface reaction has a value of 1.5×10^{-5} mol·s·kg^{-1}·m^{-1} [18].

Figure 5 shows the result of simulations of a time-dependent surface reaction. On top, a plot that corresponds to the amount of ZnO film deposited as a result of CVD taking place is shown. Under each plot, a 2D color plot of the CVD reaction rate can be seen. It is clear that the highest value of reaction rate R_{AB} would lead to a thicker, CVD regime deposition. Confirming previous simulations described above, as the deposition gap increases, the diffusion of reactants presents more "overlap" and hence, the reaction rate is higher, leading to a higher CVD component in the process, which in turn yields higher surface concentration as the gap is increased (Figure 5a).

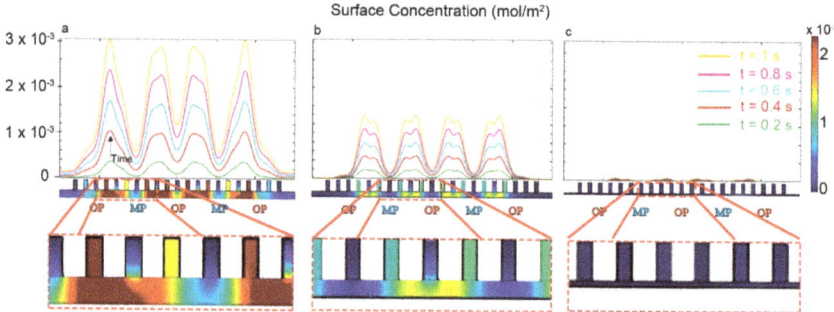

Figure 5. Results of the CVD surface reaction on the substrate calculated with a time-dependent multiphysics simulation. The plots shown correspond to the surface concentration that results from different gaps. Under each plot, a surface plot of the CVD reaction rate that corresponds to the plot directly above is shown, with OP and MP representing the outlets of the reactants; (**a**) deposition gap of 750 µm, (**b**) deposition gap of 425 µm, and (**c**) deposition gap of 150 µm.

To obtain evidence of the existence of the CVD and ALD regime with a simple change in gap, depositions of ZnO were made at different values of the gap. Figure 6 shows experimental results as evidence of the ability to modify the growth regime in our SALD system. Figure 6a presents the increase of growth rate as the gap value increases. The values for growth rate are in accordance with those reported for a self-limited (ALD) growth of ZnO [20–22]. The increase of the growth per cycle (GPC) with the gap value confirms the transition from an ALD regime (with small gap values) to a CVD regime (with higher gap values). Figure 6b shows the XRD spectra of ZnO films grown with different gap values for the same number of cycles. The peaks correspond to those of crystalline ZnO and one can observe that, as the gap value increases, the intensity of the peaks increases as well, indicating that thicker films are obtained in the same deposition time as the gap is increased.

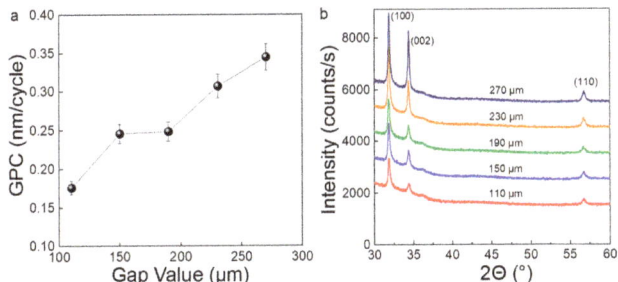

Figure 6. Experimental results for a deposition of ZnO using di-ethyl zinc (DEZ) and water as co-reactants. (**a**) Growth per cycle (GPC) evolution with different gap values; (**b**) X-ray diffraction patterns for ZnO films grown with different gap values, showing the crystalline peaks corresponding to wurzite ZnO (ICSD #82028).

3.3. Efficiency of the Deposition System Exhaust

Using the same method described above, the exhaust efficiency was studied. In the geometry of our SALD deposition head, one assumes that all inputs will be directed towards the exhausts and that all surplus of reactant concentration is directed towards the exhausts (empty arrows in Figure 2). Nevertheless, to ensure this, the exhaust outlet must have the same outflow rate as the sum of all the inflow of gases (i.e., mass balance). Failure to achieve this, i.e., due to a bad design or to a partial or total blockage of the exhausts, may induce a CVD regime even with a small deposition gap. We define exhaust efficiency as the efficiency in which the incoming gaseous reactants and by-products are extracted from the reaction zone. A high exhaust efficiency may be achieved by a properly designed outlet/exhaust area ratio, or alternatively with, for example, properly chosen pumping in the exhausts. Here, we use a geometrical approach to such efficiency by calculating the outlet/exhaust area ratio, as explained below.

The result of simulations is shown for different exhaust efficiencies in Figure 7. We quantify an exhaust efficiency as the ratio between the total cross-section area of the exhausts and the total cross-section area of the gas outlets. An ideal ratio would be where the exhaust and the outlet have the same total area, in which case we consider an exhaust efficiency of 100%. As the exhaust efficiency decreases, there is more diffusion of the reactant concentrations, even when the deposition gap is at a "close-proximity" of 150 µm. Interestingly, the diffusion of reactant concentrations has a slightly different behavior in this case than in the case of an increase of deposition gap. As the exhaust efficiency decreases, the CVD reaction rate appears to be more localized. This leads to a localized deposition in the CVD regime, and with time, four overlapped regions appear in localized points over the substrate.

To evaluate the behavior of the exhaust efficiency in real conditions, an experimental "static deposition" was performed to observe the behavior of the real SALD. The deposition head was placed at 150 µm and all flows (precursor, oxidant and separation nitrogen) were injected as usual. The movement of the substrate was suppressed, and the pattern deposited was compared to simulations.

Figure 8 shows a "static deposition" using DEZ as metallic precursor and H_2O as the oxidant. The deposited sample shows four well-defined stripes roughly at the location under the oxidant, similar to the plot of the exhaust efficiency of 4.5% shown in Figure 7. The exhaust/outlet ratio measured on the physical exhausts and outlets of the geometry of our SALD system (measured as the ratio of the total cross-section area available for exhaust and the total cross-section area of gas outlets in our deposition head) is 10.1%, which may explain the four lines pattern observed in our "static deposition". In order to improve the exhaust efficiency, one could envision either a change of the geometry of the deposition head or a forced exhaust using a pump. While it should be noted that adding a pump to the system will affect the fluid dynamics in the system, it may substantially change the deposition.

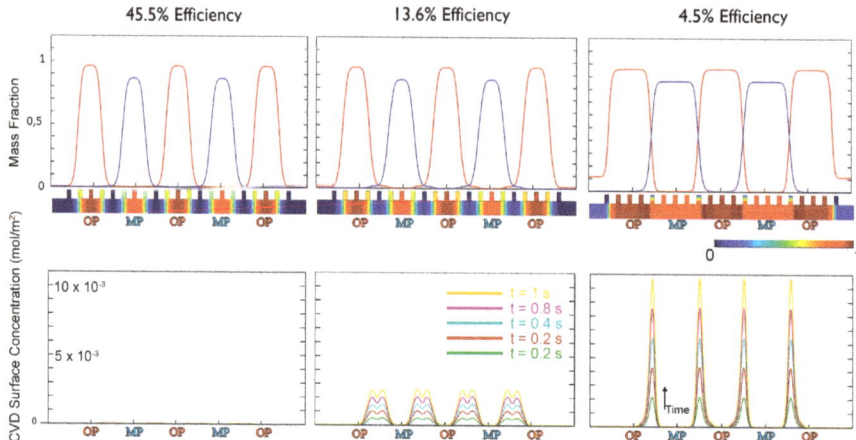

Figure 7. CVD regime deposition with different exhaust efficiencies. It can be seen that the exhaust efficiency has a drastic influence on the appearance of CVD regime: with an exhaust efficiency of ~45%, almost no appearance of CVD regime can be observed, at ~13% CVD regime appears in some regions, and at ~4% CVD regime is more pronounced and more localized.

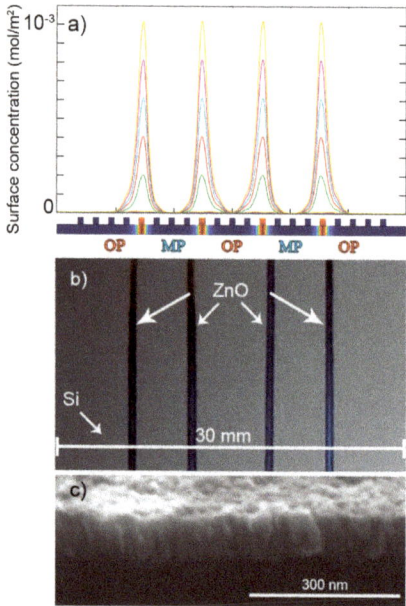

Figure 8. Simulation and experimental result of a static deposition experiment made with the SALD set-up at LMGP: (**a**) CVD deposition simulated for an exhaust efficiency of ~4% shown previously in Figure 7, and (**b**) optical photography of the pattern obtained after performing an experimental "static deposition" on a Si substrate with DEZ and H_2O; (**c**) Scanning electron microscope (SEM) cross-section image showing a ZnO thickness of ~75 nm for one of the lines in the pattern obtained after a 30 s long "static deposition".

3.4. CVD Regime Influenced by a Tilt in the Deposition Gap

Finally, the CVD surface reaction was used to assess the influence of a tilt on either the substrate or the head. Since our SALD system uses a "close-proximity" approach where the deposition gap should be around 150 µm, a slight misalignment is expected to affect the distribution of gases (flows and pressures) and thus the deposition and homogeneity of the films. Figure 9 shows a CVD surface reaction of a tilted substrate by 0.3°.

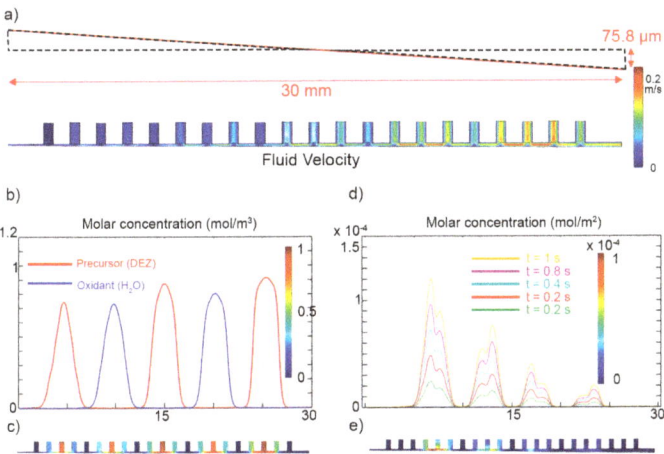

Figure 9. Simulation results for a tilt of 0.3°: (**a**) Schematic of the rotation of the substrate with respect to the deposition head, about its center point, which leads to a difference of ±75.8 µm on each side of the substrate for a 30 mm length head. Below, the velocity of the flow is shown; (**b**) concentration of reactants above the surface; and (**c**) 2D concentration distribution in the gap. The concentration slightly favors the section with a higher deposition gap, (**d**) surface concentration of a film deposited as a consequence of the appearance of a CVD regime, and (**e**) 2D distribution of the CVD reaction rate.

As it can be seen in Figure 9, the influence of a small tilt can greatly affect the deposition. The 0.3° tilt along the middle of the substrate for a 30 mm long deposition head causes a difference of ~75.8 µm at each extreme of the substrate, which changes the effective deposition gap, causing the deposition regime to change locally. Nevertheless, counter-intuitively to what was expected, the CVD reaction rate is higher in the section in which the deposition gap is narrower. This may be explained by the fact that, as the deposition gap decreases, the velocity, and hence the diffusion of the reactants, becomes higher, and the reactants have a higher chance of interacting between them creating a higher CVD reaction rate. As the gap widens, the velocity decreases and the flows are better separated by the nitrogen separation line and the exhausts, reducing the concentration of reactants and therefore the CVD reaction rate.

4. Conclusions

In the present work, Comsol Multiphysics® was used to study the influence of several geometrical parameters on the deposition regime when using a close-proximity SALD system. We confirmed that the deposition gap is crucial for the determination of an ALD or a CVD deposition regime and studied the influence of the "close-proximity" approach in the fluid dynamics. For a large deposition gap (i.e., larger than ~500 µm), the separation of the flows is not achieved and intermixing of gas concentration appears. CVD regime can, therefore, be tuned simply by changing the value of the deposition gap. Hence, we present a simple and versatile way to tune the SALD deposition process by controlling simple parameters such as the head-substrate gap in the system.

With respect to the CVD surface reaction, simulations were performed to study the CVD reaction rate at the surface of the substrate. Using the intermixing of concentrations calculated, two parameters were considered to determine the CVD reaction rate, and therefore, the CVD surface film formation, namely, the deposition gap and the exhaust efficiency. As expected, as the deposition gap is increased, the CVD reaction rate is also increased, and it gives place to a non-self-limited surface reaction on the substrate. In a 1 s simulation, a surface concentration of a film deposited in a CVD regime was plotted and a higher surface concentration was observed with a higher deposition gap. Concerning the exhaust efficiency, the ratio between the cross-section area of the exhausts with respect to the cross-section area of the outlets was investigated since this would define the exhaust efficiency for the surplus of precursor (and by-products of the surface reaction) present at the moment of deposition. Interestingly, with a ratio of 4.5%, well localized intermixing of reactants was formed, leading to a four-stripes deposition pattern that is in accordance with the deposition obtained in a static experiment of a "static-deposition" performed. This would indicate that probably the exhaust efficiency of the current deposition head geometry must consider an exhaust/outlet ratio of less than 10%, which is a value consistent with the real geometry of the deposition head used experimentally. Furthermore, with a precise mechanical design (of the gap and the inlets and exhausts of the deposition head) and with an optimized exhaust pumping, this behavior can be exploited as an approach for a selective area CVD or ALD deposition of materials.

Finally, the influence on a head-substrate tilt was studied. This is important since mis-alignment between the deposition head and the substrate is often difficult to avoid, and doing as such would require high-precision equipment, increasing the complexity of the instrumentation. Simulations show that a slight tilt of only 0.3° is enough to change the deposition regime from one side of the substrate to the other. The simulations performed regarding the tilt influence have the aim to show the sensibility of the SALD to the positioning and portray the need of high precision on the system to precisely control the deposition.

The simulations performed in this study justify the need of high-precision and alignment of the geometry used. The parallelism of the system needs to be ensured as much as possible so that the deposition regime can be well controlled. In the current study, no movement of the substrate was studied but the same approach can be readily applied to evaluate the influence of the movement on the deposition regime, as well as on the precision requirements of the geometry.

As observed here, the geometry of the system is crucial to understand and control the deposition of thin films with SALD. With the correct geometry, area-selective deposition of films with properties comparable to ALD deposited films is possible, creating new advantages for SALD such as patterned mask-less film deposition, subsequent depositions for a multi-layer scheme, or a deposition with well optimized flows and concentrations to optimize the usage of ALD precursors.

Author Contributions: Conceptualization, C.M.d.l.H. and V.H.N.; Methodology, C.M.d.l.H. and V.H.N.; Validation, J.-M.D., C.J. and D.M.-R.; Investigation, C.M.d.l.H. and V.H.N.; Writing–Original Draft, C.M.H.; Writing–Review & Editing, C.M.d.l.H., V.H.N., J.-M.D., D.B., C.J. and D.M.-R.; Supervision, D.B., C.J. and D.M.-R.; Funding Acquisition, D.M.-R.

Funding: This work was benefited from funding from the Consejo Nacional de Ciencia y Tecnología (CONACYT) from Mexico (No. 456558). The authors thank the "ARC Energies Auvergne-Rhône Alpes", for their economic support through a Ph.D. grant, and the Agence Nationale de Recherche (ANR, France) via the project DESPATCH (No. ANR-16-CE05-0021). This work benefited from the facilities and expertise of the OPE)N(RA characterization platform of FMNT (FR 2542, fmnt.fr) supported by CNRS, Grenoble INP, and UGA. DMR acknowledges funding through the Marie Curie Actions (FP7/ 2007–2013, Grant No. 631111). This project was financially supported by "Carnot Energies du Futur" (ALDASH project). This project has received funding from the European Union's Horizon 2020 FETOPEN-1-2016-2017 research and innovation programme under grant agreement 801464.

Acknowledgments: The author thanks Dominique De Barros for his support in the development of the SALD system at LMGP.

Conflicts of Interest: The authors declare no conflict of interest.

References

1. Johnson, R.W.; Hultqvist, A.; Bent, S.F. A brief review of atomic layer deposition: From fundamentals to applications. *Mater. Today* **2014**, *17*, 236–246. [CrossRef]
2. Ahvenniemi, E.; Akbashev, A.R.; Ali, S.; Bechelany, M.; Berdova, M.; Boyadjiev, S.; Cameron, D.C.; Chen, R.; Chubarov, M.; Cremers, V.; et al. Review Article: Recommended reading list of early publications on atomic layer deposition—Outcome of the "Virtual Project on the History of ALD". *J. Vac. Sci. Technol. A* **2017**, *35*, 010801. [CrossRef]
3. Muñoz-Rojas, D.; MacManus-Driscoll, J. Spatial atmospheric atomic layer deposition: A new laboratory and industrial tool for low-cost photovoltaics. *Mater. Horiz.* **2014**, *1*, 314–320. [CrossRef]
4. Muñoz-Rojas, D.; Nguyen, V.H.; de la Huerta, C.M.; Aghazadehchors, S.; Jiménez, C.; Bellet, D. Spatial atomic layer deposition (SALD), an emerging tool for energy materials. Application to new-generation photovoltaic devices and transparent conductive materials. *Comptes Rendus Phys.* **2017**, *18*, 391–400. [CrossRef]
5. Levy, D.H.; Freeman, D.; Nelson, S.F.; Cowdery-Corvan, P.J.; Irving, L.M. Stable ZnO thin film transistors by fast open air atomic layer deposition. *Appl. Phys. Lett.* **2008**, *92*, 192101. [CrossRef]
6. Nelson, S.F.; Ellinger, C.R.; Levy, D.H. Improving yield and performance in ZnO thin-film transistors made using selective area deposition. *ACS Appl. Mater. Interfaces* **2015**, *7*, 2754–2759. [CrossRef] [PubMed]
7. Nguyen, V.H.; Gottlieb, U.; Valla, A.; Muñoz, D.; Bellet, D.; Muñoz-Rojas, D. Electron tunneling through grain boundaries in transparent conductive oxides and implications for electrical conductivity: The case of ZnO:Al thin films. *Mater. Horiz.* **2018**, *5*, 715–726. [CrossRef]
8. Choi, H.; Shin, S.; Jeon, H.; Choi, Y.; Kim, J.; Kim, S.; Chung, S.C.; Oh, K. Fast spatial atomic layer deposition of Al_2O_3 at low temperature (<100 °C) as a gas permeation barrier for flexible organic light-emitting diode displays. *J. Vac. Sci. Technol. A* **2016**, *34*, 01A121. [CrossRef]
9. Poodt, P.; Lankhorst, A.; Roozeboom, F.; Spee, K.; Maas, D.; Vermeer, A. High-speed spatial atomic-layer deposition of aluminum oxide layers for solar cell passivation. *Adv. Mater.* **2010**, *22*, 3564–3567. [CrossRef] [PubMed]
10. Franke, S.; Baumkötter, M.; Monka, C.; Raabe, S.; Caspary, R.; Johannes, H.H.; Kowalsky, W.; Beck, S.; Pucci, A.; Gargouri, H. Alumina films as gas barrier layers grown by spatial atomic layer deposition with trimethylaluminum and different oxygen sources. *J. Vac. Sci. Technol. A* **2016**, *35*, 01B117. [CrossRef]
11. Muñoz-Rojas, D.; Jordan, M.; Yeoh, C.; Marin, A.T.; Kursumovic, A.; Dunlop, L.; Iza, D.C.; Chen, A.; Wang, H.; MacManus-driscoll, J.L. Growth of ~ 5 cm^2 V^{-1} s^{-1} mobility, p-type Copper (I) oxide (Cu_2O) films by fast atmospheric atomic layer deposition (AALD) at 225 °C and below. *AIP Adv.* **2012**, *2*, 042179. [CrossRef]
12. Poodt, P.; Knaapen, R.; Illiberi, A.; Roozeboom, F.; van Asten, A. Low temperature and roll-to-roll spatial atomic layer deposition for flexible electronics. *J. Vac. Sci. Technol. A* **2012**, *30*, 01A142. [CrossRef]
13. Levy, D.H. Process for Atomic Layer Deposition. US Patent 7,413,982 B2, 19 August 2008.
14. Hoye, R.L.; Muñoz-Rojas, D.; Musselman, K.P.; Vaynzof, Y.; MacManus-Driscoll, J.L. Synthesis and modeling of uniform complex metal oxides by close-proximity atmospheric pressure chemical vapor deposition. *ACS Appl. Mater. Interfaces* **2015**, *7*, 10684–10694. [CrossRef] [PubMed]
15. Deng, Z.; He, W.; Duan, C.; Chen, R.; Shan, B. Mechanistic modeling study on process optimization and precursor utilization with atmospheric spatial atomic layer deposition. *J. Vac. Sci. Technol. A* **2015**, *34*, 01A108. [CrossRef]
16. Van Deelen, J.; Illiberi, A.; Kniknie, B.; Steijvers, H.; Lankhorst, A.; Simons, P. APCVD of ZnO:Al, insight and control by modeling. *Surf. Coat. Technol.* **2013**, *230*, 239–244. [CrossRef]
17. Pan, D.; Jen, T.C.; Yuan, C. Effects of gap size, temperature and pumping pressure on the fluid dynamics and chemical kinetics of in-line spatial atomic layer deposition of Al_2O_3. *Int. J. Heat Mass Transf.* **2016**, *96*, 189–198. [CrossRef]
18. Dobkin, D.M.; Zuraw, M.K. *Principles of Chemical Vapor Deposition*, 1st ed.; Springer Science & Business Media: Dordrecht, The Netherlands, 2013.
19. Yanguas-Gil, A. *Growth and Transport in Nanostructured Materials: Reactive Transport in PVD, CVD, and ALD*; Springer International Publishing: Cham, Switzerland, 2017.
20. Lim, J.; Lee, C. Effects of substrate temperature on the microstructure and photoluminescence properties of ZnO thin films prepared by atomic layer deposition. *Thin Solid Films* **2007**, *515*, 3335–3338. [CrossRef]

21. Pal, D.; Mathur, A.; Singh, A.; Singhal, J.; Chattopadhyay, S. Photoluminescence of atomic layer deposition grown ZnO nanostructures. *Mater. Today Proc.* **2018**, *5*, 9965–9971. [CrossRef]
22. Rauwel, E.; Nilsen, O.; Rauwel, P.; Walmsley, J.C.; Frogner, H.B.; Rytter, E.; Fjellvåg, H. Oxide coating of alumina nanoporous structure using ALD to produce highly porous spinel. *Chem. Vap. Depos.* **2012**, *18*, 315–325. [CrossRef]

© 2018 by the authors. Licensee MDPI, Basel, Switzerland. This article is an open access article distributed under the terms and conditions of the Creative Commons Attribution (CC BY) license (http://creativecommons.org/licenses/by/4.0/).

Article

Growth without Postannealing of Monoclinic VO$_2$ Thin Film by Atomic Layer Deposition Using VCl$_4$ as Precursor

Wen-Jen Lee * and Yong-Han Chang

Department of Applied Physics, National Pingtung University, Pingtung 90003, Taiwan; crocodile710280@gmail.com
* Correspondence: wenjenlee@mail.nptu.edu.tw; Tel.: +886-8-7663800

Received: 18 October 2018; Accepted: 26 November 2018; Published: 27 November 2018

Abstract: Vanadium dioxide (VO$_2$) is a multifunctional material with semiconductor-to-metal transition (SMT) property. Organic vanadium compounds are usually employed as ALD precursors to grow VO$_2$ films. However, the as-deposited films are reported to have amorphous structure with no significant SMT property, therefore a postannealing process is necessary for converting the amorphous VO$_2$ to crystalline VO$_2$. In this study, an inorganic vanadium tetrachloride (VCl$_4$) is used as an ALD precursor for the first time to grow VO$_2$ films. The VO$_2$ film is directly crystallized and grown on the substrate without any postannealing process. The VO$_2$ film displays significant SMT behavior, which is verified by temperature-dependent Raman spectrometer and four-point-probing system. The results demonstrate that the VCl$_4$ is suitably employed as a new ALD precursor to grow crystallized VO$_2$ films. It can be reasonably imagined that the VCl$_4$ can also be used to grow various directly crystallized vanadium oxides by controlling the ALD-process parameters.

Keywords: vanadium dioxide; atomic layer deposition; vanadium oxide; vanadium tetrachloride; semiconductor-to-metal transition

1. Introduction

Vanadium dioxide (VO$_2$) has attracted extensive research interest during the past decades owing to its unique behavior, called semiconductor-to-metal transition (SMT) or insulator-to-metal transition (IMT), which accompanies the reversible and ultrafast phase transition between monoclinic VO$_2$ [VO$_2$(M)] and tetragonal rutile VO$_2$ [VO$_2$(R)] at temperatures around 340 K (~67 °C) [1–5]. Thus, the optical and electrical properties of VO$_2$ can be switched by controlling the SMT behavior of VO$_2$ [6,7]. Numerous factors for adjusting the SMT behavior of VO$_2$ have already been established that include impurity doping [8], stoichiometry [9], strain [10], grain boundary [11], oxygen vacancy [12], applied external electrical field [13], and light irradiation [14]. Therefore, the VO$_2$ has been widely investigated as a key material for applications in the smart thermochromic windows [15], two-terminal electronic devices [16], electric-field-effect three-terminal devices [17,18], integrated optical circuits [19], electronic oscillators [20], metamaterials [21], memristive devices [22], programmable critical thermal sensors [23], gas sensors [24], and so forth.

Various techniques had been employed for preparing VO$_2$ films, including the sol–gel method [22,23], electron-beam evaporation [25,26], sputtering [5,17], pulsed laser deposition (PLD) [27,28], molecular beam epitaxy (MBE) [16,29], chemical vapor deposition (CVD) [30–34], and atomic layer deposition (ALD) [35–50]. Among them, ALD is an excellent technique which has drawn much attention due to its many advantages, including preparation of the highly conformal thin films with almost 100% step coverage, accurate control of film thickness at the atomic scale, low growth temperature, and wide-area uniformity. These features make ALD a powerful technique for the fabrication of emerging nanostructures and nanodevices [51–53].

Generally, organic vanadium compounds are employed as ALD precursors and reacted with H_2O or O_3 to grow vanadium oxide thin films, such as tetrakis(ethylmethylamino)vanadium (TEMAV) [35–42], vanadyl isopropoxide (VTIP) [43–46], and vanadyl triisopropoxide (VTOP) [47–50]. However, the organic precursors (TEMAV, VTIP, and VTOP) are only suitable for low process temperatures in ALD because the decomposition temperatures of TEMAV, VTIP, and VTOP are about 175 [35,40], 200 [43], and 180 °C [49,50], respectively. When the process temperature is higher than the decomposition temperature of the precursor, the growth mechanism of film will be changed from ALD to CVD-like mode [35,40,43,49,50]. In this case, the film is grown by CVD instead of ALD. This is why the low temperature of 120–170 °C is usually used for the film growth of ALD using TEMAV, VTIP, or VTOP as precursor. However, the low process temperature is not enough to grow crystalline films. Therefore, the as-deposited vanadium oxide films grown by ALD from organic vanadium precursors are generally reported to have amorphous structures with no significant SMT behavior and a postannealing process is necessary for converting the amorphous to crystalline VO_2. Previous studies have reported that postannealing in N_2, He, O_2, or N_2/O_2 mixed gas with a low O_2 partial pressure resulted in crystalline monoclinic VO_2 for the temperature range of 425–585 °C [35–42]. Since the extra postannealing process is necessary to obtain a crystalline VO_2, it results in higher manufacturing costs and increases the failure possibilities of the process and products.

This work reports that a directly crystalline VO_2 film has been successfully grown by ALD using vanadium tetrachloride (VCl_4) and H_2O as precursors at a reaction temperature of 350 °C without any postannealing process. It is noticed that the inorganic VCl_4 is used as an ALD precursor for the first time, although a few papers reported that the VCl_4 had been used in traditional chemical vapor deposition (CVD) techniques [30–32]. The VO_2 film has a significant SMT behavior with a $VO_2(M)$-to-$VO_2(R)$ phase-transition temperature of about 61 ± 1 °C, which is verified from temperature-dependent Raman spectra and sheet-resistance variations of VO_2 film. Besides, the VO_2 film exhibits two orders of magnitude change in sheet resistance across the semiconductor-to-metal transition although the film thickness is only 30 nm. The results demonstrated that crystalline VO_2 films can be directly grown by ALD using VCl_4 and H_2O as precursors without any postannealing process, presenting a new selection of precursors for the ALD process to grow the crystalline VO_2 films.

2. Materials and Methods

In this work, VO_2 films were grown on native silicon-oxide-covered Si (100) substrates by ALD at 350 °C with 1000 reaction cycles. VCl_4 and H_2O were employed as precursors to grow the VO_2 films, and Ar was used as purge gas. The reservoirs of the VCl_4 and H_2O precursors were kept at the temperatures of 30 and 25 °C, respectively. The dosing rates of VCl_4 and H_2O were 0.288 and 0.296 cc/pulse, respectively, as determined by the reservoir temperature and vapor injection time. The flow rate of Ar was 5 sccm, as controlled by a mass flow controller (MFC, SEC-4400M, HORIBA STEC, Kyoto, Japan). An eight-step sequence of gas injection was applied in an ALD cycle, as combined four conventional gas-injection steps and four additional pump-down steps. The pump-down steps can effectively evacuate excess precursors and byproducts to obtain high-quality films with low Cl impurity contents and ensure the achievement of "true ALD mode" growth [54,55]. The time for each step in an ALD cycle was 0.1, 1, 0.5, 1, 0.5, 1, 0.5, and 1 s for VCl_4 vapor injection, pump-down, Ar purge, pump-down, H_2O vapor injection, pump-down, Ar purge, and pump-down, respectively.

The crystalline structures of the VO_2 films were examined by an X-ray diffractometer (XRD, D8 Advance Eco, Bruker, Karlsruhe, Germany) at 30 and 90 °C. The surface morphologies of the film were observed with a high-resolution scanning electron microscope (SEM, SU8000, Hitachi, Tokyo, Japan). In addition, in order to obtain real surface morphology of the film, the SEM analysis was performed without any conductive coating on the VO_2 film surface. The cross-sectional microstructures of the VO_2

films were observed by a high-resolution transmission electron microscope (TEM, JEM-2100F, JEOL, Tokyo, Japan). The film thickness was measured from the cross-sectional TEM micrograph and the growth rate of the VO$_2$ film was estimated from an equation of "growth rate = (film thickness/numbers of ALD cycles)". The chemical composition of the VO$_2$ film was analyzed by a high-resolution X-ray photoelectron spectrometer (XPS, Quantera SXM, ULVAC-PHI, Chigasaki, Japan). In addition, the XPS analysis was performed on the VO$_2$ film surface before and after Ar ion etching with an etching depth of about 2 nm. The temperature-dependent Raman spectra of the VO$_2$ film were examined at temperatures between 30 and 80 °C by a micro Raman spectrometer (Raman, UniRAM II, Uninanotech, Yongin, Korea) with a temperature-controllable sample stage. The temperature-dependent sheet resistance of the VO$_2$ film was measured at temperatures between 30 and 90 °C by a Keithley 2614B SourceMeter (Keithley, Solon, OH, USA) under a four-point probing configuration with a temperature-controllable sample stage.

3. Results and Discussion

Figure 1 shows the XRD patterns of VO$_2$ film measured at 30 and 90 °C, which demonstrate the structural transition of VO$_2$ film from monoclinic (30 °C) to tetragonal rutile (90 °C) phase. The VO$_2$ film measured at 30 °C (Figure 1a) displays that two XRD peaks located at 2θ of 27.9° and 55.4° can be indexed to the (011) and (220) planes of monoclinic VO$_2$(M) (JCPDS no.: 82–0661), respectively. When the temperature is raised to 90 °C (Figure 1b), two XRD peaks located at 2θ of 27.6° and 55.4° are detected, which can be assigned to the (110) and (211) planes of tetragonal rutile VO$_2$(R) (JCPDS no.: 79–1655), respectively. In addition, because the VO$_2$ film is grown on Si substrate, an obvious XRD peak of Si (113) at 2θ of 51.9° has also been detected (Supplementary materials: Part 1). Figure 1c is a comparison of Figure 1a,b, which clearly shows an XRD peak shift of VO$_2$(M) (011) to VO$_2$(R) (110) peak in 26° ≤ 2θ ≤ 30° and the XRD peak of Si substrate does not shift. The XRD peak-shifting behavior is a diagnostic feature for the phase transition of VO$_2$ film from monoclinic to tetragonal rutile structure. Previously, the similar XRD peak-shifting phenomenon had also been reported by Wu et al. [56] for confirming the phase transformation of monoclinic VO$_2$ to tetragonal rutile VO$_2$.

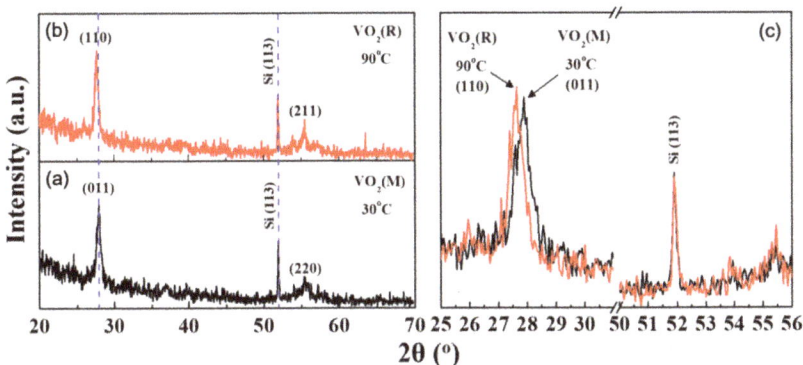

Figure 1. XRD patterns of VO$_2$ film measured at (a) 30 °C and (b) 90 °C. (c) A comparison of (a) and (b).

Figure 2 shows the SEM and TEM analyzed results for surface and cross-sectional microstructures of the VO$_2$ film, respectively. The SEM image of surface morphology of the VO$_2$ film (Figure 2a) displays a conformal VO$_2$ film with bigger grains surrounded by small grains; the grain sizes of big and small grains are about 78 ± 14 and 40 ± 6 nm, respectively. According to the cross-sectional TEM bright-field and dark-field images (Figure 2b,c), it can be clearly observed that the VO$_2$ film is grown on a native oxide layer (SiO$_x$) of Si substrate and constructed from columnar grains. The thickness of the VO$_2$ film is about 30 nm, and displays its growth rate at about 0.03 nm/cycle. Moreover, the VO$_2$

grains are directly crystallized and grown on the top surface of the native oxide layer, which is verified by the high-resolution TEM (HR-TEM) image of the $VO_2/SiO_x/Si$ interface (Figure 2d). The HR-TEM image also reveals a clear lattice fringe of about 0.32 nm, corresponding to the (011) plane of VO_2(M). The selected-area electron-beam diffraction (SA-EBD) pattern obtained by focusing the electron beam on an individual VO_2 grain is shown in Figure 2e; the SA-EBD pattern can be indexed to monoclinic VO_2(M) in agreement with the XRD results.

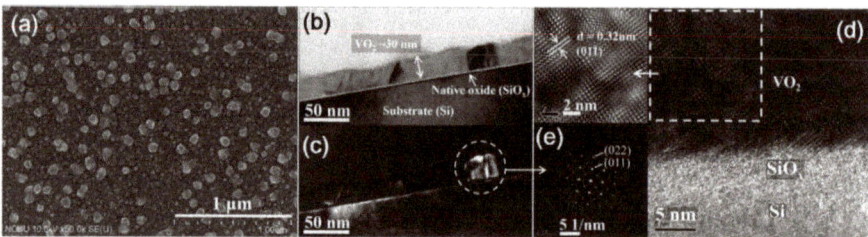

Figure 2. Microstructure analyses of VO_2 film. (**a**) An SEM top-view image. TEM cross-sectional (**b**) bright-field and (**c**) dark-field images. (**d**) A high-resolution TEM (HR-TEM) image of the $VO_2/SiO_x/Si$ interface. (**e**) The selected-area electron-beam diffraction (SA-EBD) pattern obtained by focusing the electron beam on an individual VO_2 grain.

Figure 3 shows the V2p, O1s, and Cl2p XPS spectra for original (before etching) and after argon-ion etching surface of VO_2 film. In addition, the chemical composition of VO_2 film calculated from XPS spectra are shown in Table 1. The V and O concentrations of VO_2 film are about 25.7 at.% and 74.3 at.%, respectively, for the film before surface etching and are about 33.1 at.% and 66.9 at.%, respectively, for the film after surface etching. The results clearly indicate that the original surface of the VO_2 film has a higher oxygen concentration because the VO_2 film was exposed to air (oxygen-rich) environment, resulting in absorption of oxygen and a native oxide layer (overoxidation layer) forming on the surface of the VO_2 film [26,33,34,39,41,43,57,58]. After argon-ion etching, surface contamination and the native oxide layer of the VO_2 film had been removed, and the atomic proportion of V:O atom was about 1:2 in agreement with the stoichiometry of VO_2. Besides, no Cl impurity had been detected in the VO_2 film, demonstrating the Cl concentration in the VO_2 film was lower than the detection limit of XPS (approximately 0.1 at.%). It is noteworthy that this work successfully achieved VO_2 film with high purity (Cl impurity <0.1 at.%) by using a low growth temperature of 350 °C, which can be attributed to the additional pump-down steps in the ALD reaction cycles effectively evacuating excess precursors and byproducts [54]. In a previous study, Cheng et al. reported that implementation of pump-down steps into the gaseous-pulse cycle of ALD can effectively reduce the Cl residues. They used $TiCl_4$ as ALD precursor to grow TiN films by using conventional four-step ALD and modified six-step ALD (adding two pump-down steps). Their results showed that the Cl residues of TiN films can be decreased from about 7.7 at.% to 2.3 at.% at the growth temperature of 300 °C [54].

In Figure 3a, the V2p3/2 peak of the original VO_2 film (before surface etching) can be fitted with two peaks at binding energy of about 517.2 and 515.6 eV, which can be assigned to V^{5+} and V^{4+}, respectively [26,33,34,39,41,43,57,58]. Musschoot et al. [43] and Sliversmit et al. [57] reported that the V^{5+} signal is mainly contributed from the native oxide layer (overoxidation layer) of VO_2 film. After surface etching (to remove the native oxide layer), the $V2p_{3/2}$ peak has a maximum at 515.6 eV (assigned to V^{4+}), which primarily confirms VO_2 stoichiometry. In Figure 3b, the O1s XPS peaks are located at binding energy of about 529.8 and 530.5 eV for the original and after-surface-etching VO_2 film, showing a peak-shifting phenomenon. The similar peak-shifting phenomenon of O1s XPS peak for VO_2 film after surface etching by argon ion sputtering had also been observed by Musschoot et al. [43].

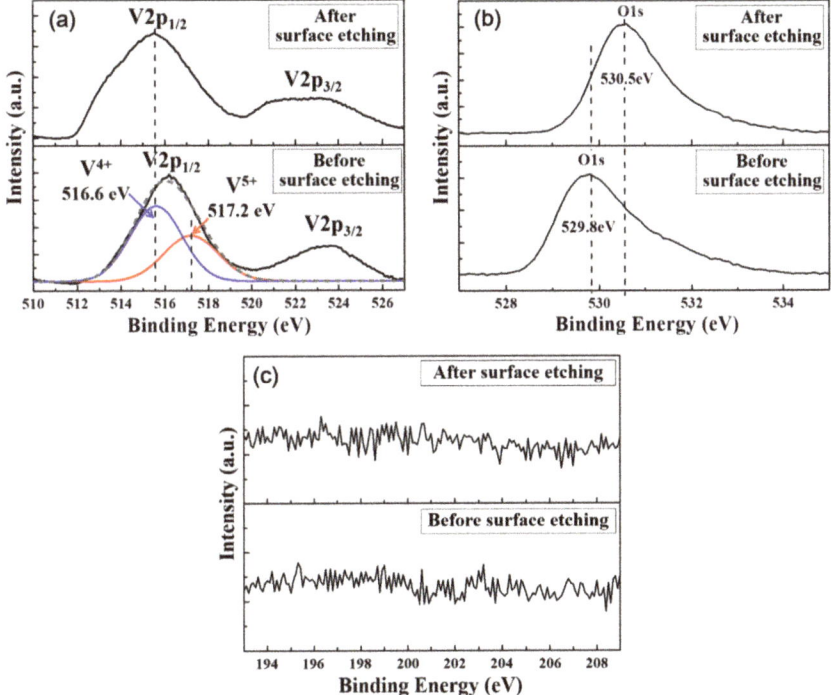

Figure 3. (a) V2p, (b) O1s, and (c) Cl2p XPS spectra for original (before etching) and after argon-ion etching surface of VO$_2$ film.

Table 1. Chemical composition of VO$_2$ thin film analyzed by XPS.

Elemental Content	Before Surface Etching	After Surface Etching
V (at.%)	25.7	33.1
O (at.%)	74.3	66.9
Cl (at.%)	<0.1	<0.1

Figure 4a,b show the selected temperature-dependent Raman spectra of the VO$_2$ film for heating and cooling cycles, respectively. It is noticed that the full temperature-dependent Raman spectra of the VO$_2$ films for temperatures between 30 and 80 °C in heating and cooling cycles are shown in Figures S1 and S2, respectively. As shown in Figure 4, four Raman peaks at 194, 224, 305, and 616 cm^{-1} are associated with the monoclinic phase VO$_2$ [25,26,59–61]. The peaks of 194, 305, and 616 cm^{-1} are assigned to A_g phonon vibration modes [25,26,59,60] and the peak of 224 cm^{-1} can be assigned to A_g + B_g mode [61]. The low-frequency phonons at 194 and 224 cm^{-1} relate to lattice motion involving V–V bonds, while the other peaks are attributed to V–O bonds [26,59–61]. Peaks located at 301, 520, and 935–990 cm^{-1} are contributed from the silicon substrate that compared with the Raman spectrum of the silicon substrate (Figure 5). Moreover, the phonon intensities of 194, 224, and 616 cm^{-1} gradually disappear as the temperature increases and display the reversibility during the cooling cycle. However, the peak intensity of 305 cm^{-1} does not show an evident change due to an overlap signal between 305 and 301 cm^{-1} for VO$_2$ and silicon substrate, respectively.

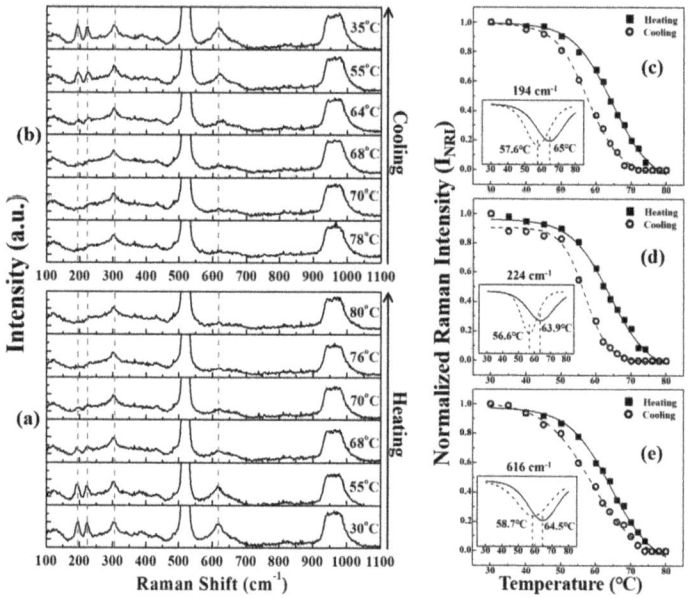

Figure 4. Temperature-dependent Raman spectra of the VO$_2$ film: (**a**) heating cycle, (**b**) cooling cycle, and relative Raman intensity of the A$_g$ phonon mode at (**c**) 194, (**d**) 224, and (**e**) 616 cm^{-1}.

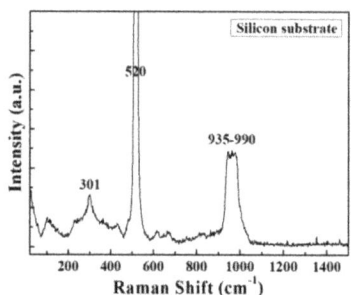

Figure 5. Raman spectrum of the silicon substrate.

Furthermore, the plots of normalized Raman intensity variations for the A$_g$ phonon vibration mode at 194, 224, and 616 cm^{-1} are shown in Figure 4c,d,e, respectively. The normalized Raman intensity of VO$_2$ film was calculated from the equation below:

$$I_{NRI} = \frac{I_T - I_{80}}{I_{30} - I_{80}} \quad (1)$$

where I_{NRI} is the normalized Raman intensity, I_T is the Raman intensity measured at indicated temperature (T), I_{30} and I_{80} are the Raman intensities measured at 30 and 80 °C, respectively. It can be seen clearly that the plots of Raman intensity vibrations show a hysteresis feature for Raman shift at 194, 224, and 616 cm^{-1}. The phase transition temperatures of VO$_2$ film estimated by the differential curves (as inserts) are about 65, 63.9, and 64.5 °C for 194, 224, and 616 cm^{-1} in the heating process, respectively. In the cooling process, the phase transition temperatures of VO$_2$ film are about 57.6, 56.6, and 58.7 °C for 194, 224, and 616 cm^{-1}, respectively. Therefore, the overall SMT temperatures estimated from the middle of the hysteresis curves are about 61.3, 60.25, and 61.6 °C for 194, 224, and 616 cm^{-1}, respectively.

The temperature-dependent sheet-resistance (SR) variation of VO$_2$ film is shown in Figure 6, displaying a thermal hysteresis variation. Besides, the SR variation has approached two orders of magnitude across the semiconductor-to-metal transition (SMT) of the VO$_2$ film (SR changed from 2.2×10^4 to 2.7×10^2 Ω/□ for the temperature raised from 40 to 80 °C) so that the value of the resistance ratio agrees with the typical VO$_2$ film thickness less than 50 nm (typically, the resistance ratios of most VO$_2$ films across the SMT are in the range of 10^2–10^3 for thickness <50 nm) [29]. Furthermore, a sharp drop of SR can be clearly observed in the heating cycle, determining a phase transition temperature of about 63 °C, and a sharp rise of SR in the cooling cycle with a phase transition temperature of about 56 °C can be also seen in Figure 6. Therefore, the SMT temperature estimated from the middle of thermal hysteresis SR variation is about 60.0 °C.

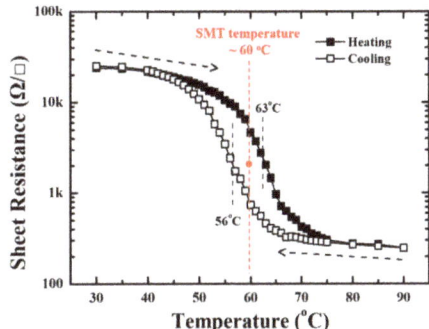

Figure 6. Temperature-dependent sheet-resistance variation of the VO$_2$ film.

There are several parameters that may affect the temperature-dependent electrical properties of VO$_2$, such as changes in impurity content, stoichiometry, strain, oxygen vacancies, and the presence of grain boundaries [8–12]. In this work, the SMT temperature of VO$_2$ film evaluated from the temperature-dependent Raman spectra and sheet-resistance variation is about 61 ± 1 °C, slightly different from the well-known 340K (~67 °C), which can be reasonably attributed to the influence of grain boundary density because the VO$_2$ film has a polycrystalline structure with considerable grain boundaries.

4. Conclusions

In conclusion, the VCl$_4$ is successfully employed as a new ALD precursor to grow a VO$_2$ film on the Si (100) substrate. Besides, without any postannealing process required, the as-deposited VO$_2$ film is directly crystallized and provides a significant SMT property. Moreover, it can be reasonably imagined that the VCl$_4$ can be used not only to grow crystalline VO$_2$ films, but also to grow other different vanadium oxides (VO$_x$, $x \neq 2$) by controlling the ALD-process parameters (such as process temperature, VCl$_4$/H$_2$O ratio, and so forth). It is just like this that the VCl$_4$ had been used as a precursor for atmospheric pressure CVD to grow different vanadium oxides (VO$_2$ and V$_2$O$_5$) by controlling process parameters of temperature and VCl$_4$/H$_2$O ratio [30–32]. We anticipate this work to be a starting point for using VCl$_4$ as a precursor to grow various directly crystallized vanadium oxides by ALD without any postannealing process.

Supplementary Materials: The following are available online at http://www.mdpi.com/2079-6412/8/12/431/s1, Part 1: A brief report provided by Bruker Corporation for explaining "Why Si (113) peak appears in GIXRD profile?" Part 2: Full temperature-dependent Raman spectra of the VO$_2$ films. Figure S1: Raman spectra of the VO$_2$ film measured at indicant temperature for heating cycle (temperature from 30 to 80 °C), Figure S2: Raman spectra of the VO$_2$ film measured at indicant temperature for cooling cycle (temperature from 78 to 35 °C).

Funding: This research was funded by the Ministry of Science and Technology of Taiwan (Nos.: MOST 105-2221-E-153-001 and MOST 106-2221-E-153-004).

Acknowledgments: The authors would like to thank Dr. Ya-Ching Yang (Bruker Corporation) for discussing and providing a brief report to explain "Why Si (113) peak appears in GIXRD profile?", Ms. Hui-Jung Shih (Instrument Center, NCKU) for HR-SEM analysis, Ms. Mei-Lan Liang and Ms. Shih-Wen Tseng (Center for Micro/Nano Science and Technology, NCKU) for FIB and TEM analyses, and Ms. Swee-Lan Cheah (Instrument Center, NTHU) for HR-XPS analysis.

Conflicts of Interest: The authors declare no conflict of interest.

References

1. Morin, F.J. Oxides which show a metal-to-insulator transition at the Neel temperature. *Phys. Rev. Lett.* **1959**, *3*, 34–36. [CrossRef]
2. Barker, A.S.; Verleur, H.W.; Guggenheim, H.J. Infrared optical properties of vanadium dioxide above and below the transition temperature. *Phys. Rev. Lett.* **1966**, *17*, 1286–1289. [CrossRef]
3. Park, J.H.; Coy, J.M.; Kasirga, T.S.; Huang, C.; Fei, Z.; Hunter, S.; Cobdem, D.H. Measurement of a solid-state triple point at the metal-insulator transition in VO_2. *Nature* **2013**, *500*, 431–434. [CrossRef] [PubMed]
4. O'Callahan, B.T.; Jones, A.C.; Park, J.H.; Cobden, D.H.; Atkin, J.M.; Raschke, M.B. Inhomogeneity of the ultrafast insulator-to-metal transition dynamics of VO_2. *Nat. Commun.* **2015**, *6*, 6849. [CrossRef] [PubMed]
5. Cueff, S.; Li, D.; Zhou, Y.; Wong, F.J.; Kurvits, J.A.; Ramanathan, S.; Zia, R. Dynamic control of light emission faster than the lifetime limit using VO_2 phase-change. *Nat. Commun.* **2015**, *6*, 8636. [CrossRef] [PubMed]
6. Yang, Z.; Ko, C.; Ramanathan, S. Oxide electronics utilizing ultrafast metal-insulator transitions. *Annu. Rev. Mater. Res.* **2011**, *41*, 337–367. [CrossRef]
7. Nakano, M.; Shibuya, K.; Okuyama, D.; Hatano, T.; Ono, S.; Kawasaki, M.; Iwasa, Y.; Tokura, Y. Collective bulk carrier delocalization driven by electrostatic surface charge accumulation. *Nature* **2012**, *487*, 459–462. [CrossRef] [PubMed]
8. Wang, N.; Liu, S.; Zeng, X.T.; Magdassi, S.; Long, Y. Mg/W-codoped vanadium dioxide thin films with enhanced visible transmittance and low phase transition temperature. *J. Mater. Chem. C* **2015**, *3*, 6771–6777. [CrossRef]
9. Zhang, S.; Kim, I.S.; Lauhon, L.J. Stoichiometry engineering of monoclinic to rutile phase transition in suspended single crystalline vanadium dioxide nanobeams. *Nano Lett.* **2011**, *11*, 1443–1447. [CrossRef] [PubMed]
10. Petraru, A.; Soni, R.; Kohlstedt, H. Voltage controlled biaxial strain in VO_2 films grown on $0.72Pb(Mg_{1/3}Nb_{2/3})$-$0.28PbTiO_3$ crystals and its effect on the transition temperature. *Appl. Phys. Lett.* **2014**, *105*, 092902. [CrossRef]
11. Jian, J.; Chen, A.; Zhang, W.; Wang, H. Sharp semiconductor-to-metal transition of VO_2 thin films on glass substrates. *J. Appl. Phys.* **2013**, *114*, 244301. [CrossRef]
12. Jeong, J.; Aetukuri, N.; Graf, T.; Schladt, T.D.; Samant, M.G.; Parkin, S.S.P. Suppression of metal-insulator transition in VO_2 by electric field-induced oxygen vacancy formation. *Science* **2013**, *339*, 1402–1405. [CrossRef] [PubMed]
13. Aetukuri, N.B.; Gary, A.X.; Drouard, M.; Cossale, M.; Gao, L.; Reid, A.H.; Kukreja, R.; Ohldag, H.; Jenkins, C.A.; Arenholz, E.; et al. Control of the metal–insulator transition in vanadium dioxide by modifying orbital occupancy. *Nat. Phys.* **2013**, *9*, 661–666. [CrossRef]
14. Wu, J.M.; Liou, L.B. Room temperature photo-induced phase transitions of VO_2 nanodevices. *J. Mater. Chem.* **2011**, *21*, 5499–5504. [CrossRef]
15. Xu, F.; Cao, X.; Luo, H.; Jin, P. Recent advances in VO_2-based thermochromic composites for smart windows. *J. Mater. Chem. C* **2018**, *6*, 1903–1919. [CrossRef]
16. Shukla, N.; Parihar, A.; Freeman, E.; Paik, H.; Stone, G.; Narayanan, V.; Wen, H.; Cai, Z.; Gopalan, V.; Engel-Herbert, R.; et al. Synchronized charge oscillations in correlated electron systems. *Sci. Rep.* **2014**, *4*, 4964. [CrossRef]
17. Ruzmetov, D.; Gopalakrishnan, G.; Ko, C.; Narayanamurti, V.; Ramanathan, S. Three-terminal field effect devices utilizing thin film vanadium oxide as the channel layer. *J. Appl. Phys.* **2010**, *107*, 114516. [CrossRef]
18. Yajima, T.; Nishimura, T.; Toriumi, A. Positive-bias gate-controlled metal–insulator transition in ultrathin VO_2 channels with TiO_2 gate dielectrics. *Nat. Commun.* **2015**, *6*, 10104. [CrossRef] [PubMed]

19. Briggs, R.M.; Pryce, I.M.; Atwater, H.A. Compact silicon photonic waveguide modulator based on the vanadium dioxide metal-insulator phase transition. *Opt. Express* **2010**, *18*, 11192–11201. [CrossRef] [PubMed]
20. Gu, Q.; Falk, A.; Wu, J.Q.; Ouyang, L.; Park, H. Current-driven phase oscillation and domain-wall propagation in $W_xV_{1-x}O_2$ nanobeams. *Nano Lett.* **2007**, *7*, 363–366. [CrossRef] [PubMed]
21. Dicken, M.J.; Aydin, K.; Pryce, I.M.; Sweatlock, L.A.; Boyd, E.M.; Walavalkar, S.; Ma, J.; Atwater, H.A. Frequency tunable near-infrared metamaterials based on VO_2 phase transition. *Opt. Express* **2009**, *17*, 18330–18339. [CrossRef] [PubMed]
22. Driscoll, T.; Kim, H.T.; Chae, B.G.; Di Ventra, M.; Basov, D.N. Phase-transition driven memristive system. *Appl. Phys. Lett.* **2009**, *95*, 043503. [CrossRef]
23. Kim, B.J.; Lee, Y.W.; Chae, B.G.; Yun, S.J.; Oh, S.Y.; Kim, H.T. Temperature dependence of the first-order metal-insulator transition in VO_2 and programmable critical temperature sensor. *Appl. Phys. Lett.* **2007**, *90*, 023515. [CrossRef]
24. Strelcov, E.; Lilach, Y.; Kolmakov, A. Gas sensor based on metal-insulator transition in VO_2 nanowire thermistor. *Nano Lett.* **2009**, *9*, 2322–2326. [CrossRef] [PubMed]
25. Heckman, E.M.; Gonzalez, L.P.; Guha, S.; Barnes, J.O.; Carpenter, A. Electrical and optical switching properties of ion implanted VO_2 thin films. *Thin Solid Films* **2009**, *518*, 265–268. [CrossRef]
26. Ureña-Begara, F.; Crunteanu, A.; Raskin, J.P. Raman and XPS characterization of vanadium oxide thin films with temperature. *Appl. Surf. Sci.* **2017**, *403*, 717–727. [CrossRef]
27. Chiu, T.W.; Tonooka, K.; Kikuchi, N. Growth of *b*-axis oriented VO_2 thin films on glass substrates using ZnO buffer layer. *Appl. Surf. Sci.* **2010**, *256*, 6834–6837. [CrossRef]
28. Zhang, P.; Jiang, K.; Deng, Q.; You, Q.; Zhang, J.; Wu, J.; Hu, Z.; Chu, J. Manipulations from oxygen partial pressure on the higher energy electronic transition and dielectric function of VO_2 films during a metal-insulator transition process. *J. Mater. Chem. C* **2015**, *3*, 5033–5040. [CrossRef]
29. Zhang, H.T.; Zhang, L.; Mukherjee, D.; Zheng, Y.X.; Haislmaier, R.C.; Alem, N.; Engel-Herbert, R. Wafer-scale growth of VO_2 thin films using a combinatorial approach. *Nat. Commun.* **2015**, *6*, 8475. [CrossRef] [PubMed]
30. Vernardou, D.; Pemble, M.E.; Sheel, D.W. The growth of thermochromic VO_2 films on glass by atmospheric-pressure CVD: A comparative study of precursors, CVD methodology, and substrates. *Chem. Vap. Depos.* **2006**, *12*, 263–274. [CrossRef]
31. Vernardou, D.; Paterakis, P.; Drosos, H.; Spanakis, E.; Povey, I.M.; Pemble, M.E.; Koudoumas, E.; Katsarakis, N. A study of the electrochemical performance of vanadium oxide thin films grown by atmospheric pressure chemical vapour deposition. *Sol. Energy Mater. Sol. Cells* **2011**, *95*, 2842–2847. [CrossRef]
32. Vernardou, D. Using an atmospheric pressure chemical vapor deposition process for the development of V_2O_5 as an electrochromic material. *Coatings* **2017**, *7*, 24. [CrossRef]
33. Makarevich, A.M.; Sadykov, I.I.; Sharovarov, D.I.; Amelichev, V.A.; Adamenkov, A.A.; Tsymbarenko, D.M.; Plokhih, A.V.; Esaulkov, M.N.; Solyankin, P.M.; Kaul, A.R. Chemical synthesis of high quality epitaxial vanadium dioxide films with sharp electrical and optical switch properties. *J. Mater. Chem. C* **2015**, *3*, 9197–9205. [CrossRef]
34. Blackburn, B.; Powell, M.J.; Knapp, C.E.; Bear, J.C.; Carmalt, C.J.; Parkin, I.P. [{$VOCl_2(CH_2(COOEt)_2)$}$_4$] as a molecular precursor for thermochromic monoclinic VO_2 thin films and nanoparticles. *J. Mater. Chem. C* **2016**, *4*, 10453–10463. [CrossRef]
35. Rampelberg, G.; Schaekers, M.; Martens, K.; Xie, Q.; Deduytsche, D.; De Schutter, B.; Blasco, N.; Kittl, J.; Detavernier, C. Semiconductor-metal transition in thin VO_2 films grown by ozone based atomic layer deposition. *Appl. Phys. Lett.* **2011**, *98*, 162902. [CrossRef]
36. Premkumar, P.A.; Toeller, M.; Radu, I.P.; Adelmann, C.; Schaekers, M.; Meersschaut, J.; Conard, T.; Elshocht, S.V. Process study and characterization of VO_2 thin films synthesized by ALD using TEMAV and O_3 precursors. *ECS J. Solid State Sci. Technol.* **2012**, *1*, P169–P174. [CrossRef]
37. Tangirala, M.; Zhang, K.; Nminibapiel, D.; Pallem, V.; Dussarrat, C.; Cao, W.; Adam, T.N.; Johnson, C.S.; Elsayed-Ali, H.E.; Baumgart, H. Physical analysis of VO_2 films grown by atomic layer deposition and RF magnetron sputtering. *ECS J. Solid State Sci. Technol.* **2014**, *3*, N89–N94. [CrossRef]

38. Cerbu, F.; Chou, H.S.; Radu, I.P.; Martens, K.; Peter, A.P.; Afanas'ev, V.V.; Stesmans, A. Band alignment and effective work function of atomic-layer deposited VO$_2$ and V$_2$O$_5$ films on SiO$_2$ and Al$_2$O$_3$. *Phys. Status Solidi C* **2015**, *12*, 238–241. [CrossRef]
39. Zhang, K.; Tangirala, M.; Nminibapiel, D.; Cao, W.; Pallem, V.; Dussarrat, C.; Baumgart, H. Synthesis of VO$_2$ thin films by atomic layer deposition with TEMAV as precursor. *ECS Trans.* **2013**, *50*, 175–182. [CrossRef]
40. Blanquart, T.; Niinistö, J.; Gavagnin, M.; Longo, V.; Heikkilä, M.; Puukilainen, E.; Pallem, V.R.; Dussarrat, C.; Ritala, M.; Leskelä, M. Atomic layer deposition and characterization of vanadium oxide thin films. *RSC Adv.* **2013**, *3*, 1179–1185. [CrossRef]
41. Kozen, A.C.; Joress, H.; Currie, M.; Anderson, V.R.; Eddy, C.R., Jr.; Wheeler, V.D. Structural characterization of atomic layer deposited vanadium dioxide. *J. Phys. Chem. C* **2017**, *121*, 19341–19347. [CrossRef]
42. Park, H.H.; Larrabee, T.J.; Ruppalt, L.B.; Culbertson, J.C.; Prokes, S.M. Tunable electrical properties of vanadium oxide by hydrogen-plasma-treated atomic layer deposition. *ACS Omega* **2017**, *2*, 1259–1264. [CrossRef]
43. Musschoot, J.; Deduytsche, D.; Poelman, H.; Haemers, J.; Van Meirhaeghe, R.L.; Van den Berghe, S.; Detavernier, C. Comparison of thermal and plasma-enhanced ALD/CVD of vanadium pentoxide. *J. Electrochem. Soc.* **2009**, *156*, P122–P126. [CrossRef]
44. Boukhalfa, S.; Evanoff, K.; Yushin, G. Atomic layer deposition of vanadium oxide on carbon nanotubes for high-power supercapacitor electrodes. *Energy Environ. Sci.* **2012**, *5*, 6872–6879. [CrossRef]
45. Singh, T.; Wang, S.; Aslam, N.; Zhang, H.; Hoffmann-Eifert, S.; Mathur, S. Atomic layer deposition of transparent VO$_x$ thin films for resistive switching applications. *Chem. Vap. Depos.* **2014**, *20*, 291–297. [CrossRef]
46. Daubert, J.S.; Lewis, N.P.; Gotsch, H.N.; Mundy, J.Z.; Monroe, D.N.; Dickey, E.C.; Losego, M.D.; Parsons, G.N. Effect of meso- and micro-porosity in carbon electrodes on atomic layer deposition of pseudocapacitive V$_2$O$_5$ for high performance supercapacitors. *Chem. Mater.* **2015**, *27*, 6524–6534. [CrossRef]
47. Baddour-Hadjean, R.; Golabkan, V.; Pereira-Ramos, J.P.; Mantoux, A.; Lincot, D. A Raman study of the lithium insertion process in vanadium pentoxide thin films deposited by atomic layer deposition. *J. Raman Spectrosc.* **2002**, *33*, 631–638. [CrossRef]
48. Badot, J.C.; Mantoux, A.; Baffier, N.; Dubrunfaut, O.; Lincot, D. Electrical properties of V$_2$O$_5$ thin films obtained by atomic layer deposition (ALD). *J. Mater. Chem.* **2004**, *14*, 3411–3415. [CrossRef]
49. Chen, X.; Pomerantseva, E.; Banerjee, P.; Gregorczyk, K.; Ghodssi, R.; Rubloff, G. Ozone-based atomic layer deposition of crystalline V$_2$O$_5$ films for high performance electrochemical energy storage. *Chem. Mater.* **2012**, *24*, 1255–1261. [CrossRef]
50. Badot, J.C.; Ribes, S.; Yousfi, E.B.; Vivier, V.; Pereira-Ramos, J.P.; Baffier, N.; Lincot, D. Atomic layer epitaxy of vanadium oxide thin films and electrochemical behavior in presence of lithium ions. *Electrochem. Solid-State Lett.* **2000**, *3*, 485–488. [CrossRef]
51. Kim, H.; Maeng, W.J. Applications of atomic layer deposition to nanofabrication and emerging nanodevices. *Thin Solid Films* **2009**, *517*, 2563–2580. [CrossRef]
52. George, S.M. Atomic layer deposition: An overview. *Chem. Rev.* **2010**, *110*, 111–131. [CrossRef] [PubMed]
53. Gelde, L.; Cuevas, A.L.; Martínez de Yuso, M.D.V.; Benavente, J.; Vega, V.; Gonzalez, A.S.; Prida, V.M.; Hernando, B. Influence of TiO$_2$-coating layer on nanoporous alumina membranes by ALD technique. *Coatings* **2018**, *8*, 60. [CrossRef]
54. Cheng, H.E.; Lee, W.J. Properties of TiN films grown by atomic-layer chemical vapor deposition with a modified gaseous-pulse sequence. *Mater. Chem. Phys.* **2006**, *97*, 315–320. [CrossRef]
55. Lee, W.J.; Hon, M.H. Space-limited crystal growth mechanism of TiO$_2$ films by atomic layer deposition. *J. Phys. Chem. C* **2010**, *114*, 6917–6921. [CrossRef]
56. Wu, C.; Zhang, X.; Dai, J.; Yang, J.; Wu, Z.; Wei, S.; Xie, Y. Direct hydrothermal synthesis of monoclinic VO$_2$(M) single-domain nanorods on large scale displaying magnetocaloric effect. *J. Mater. Chem.* **2011**, *21*, 4509–4517. [CrossRef]
57. Silversmit, G.; Depla, D.; Poelman, H.; Martin, G.B.; De Gryse, R. Determination of the V2p XPS binding energies for different vanadium oxidation states (V^{5+} to V^{0+}). *J. Electron. Spectrosc. Relat. Phenom.* **2004**, *135*, 167–175. [CrossRef]
58. Hryha, E.; Rutqvist, E.; Nyborg, L. Stoichiometric vanadium oxides studied by XPS. *Surf. Interface Anal.* **2012**, *44*, 1022–1025. [CrossRef]

59. Yuan, X.; Zhang, W.; Zhang, P. Hole-lattice coupling and photoinduced insulator-metal transition in VO_2. *Phys. Rev. B* **2013**, *88*, 035119. [CrossRef]
60. Zaghrioui, M.; Sakai, J.; Azhan, N.H.; Su, K.; Okimura, K. Polarized Raman scattering of large crystalline domains in VO_2 films on sapphire. *Vib. Spectrosc.* **2015**, *80*, 79–85. [CrossRef]
61. Shibuya, K.; Sawa, A. Polarized Raman scattering of epitaxial vanadium dioxide films with low-temperature monoclinic phase. *J. Appl. Phys.* **2017**, *122*, 015307. [CrossRef]

© 2018 by the authors. Licensee MDPI, Basel, Switzerland. This article is an open access article distributed under the terms and conditions of the Creative Commons Attribution (CC BY) license (http://creativecommons.org/licenses/by/4.0/).

Article

Growth of Atomic Layer Deposited Ruthenium and Its Optical Properties at Short Wavelengths Using Ru(EtCp)$_2$ and Oxygen

Robert Müller [1,2], Lilit Ghazaryan [1,2], Paul Schenk [1,2], Sabrina Wolleb [2], Vivek Beladiya [1], Felix Otto [3], Norbert Kaiser [2], Andreas Tünnermann [1,2], Torsten Fritz [3] and Adriana Szeghalmi [1,2,*]

1   Institute of Applied Physics, Friedrich Schiller University Jena, Albert-Einstein-Str. 15, 07745 Jena, Germany; robertmueller89@web.de (R.M.); lilit.ghazaryan@uni-jena.de (L.G.); paul.schenk@uni-jena.de (P.S.); vivek.beladiya@uni-jena.de (V.B.); andreas.tuennermann@iof.fraunhofer.de (A.T.)
2   Center of Excellence in Photonics, Fraunhofer Institute for Applied Optics and Precision Engineering IOF, Albert-Einstein-Str. 7, 07745 Jena, Germany; sabrina-jasmin.wolleb@iof.fraunhofer.de (S.W.); norbert.kaiser@iof.fraunhofer.de (N.K.)
3   Institute of Solid State Physics, Friedrich Schiller University Jena, Helmholtzweg 5, 07743 Jena, Germany; felix.otto@uni-jena.de (F.O.); torsten.fritz@uni-jena.de (T.F.)
*   Correspondence: adriana.szeghalmi@iof.fraunhofer.de; Tel.: +49-3641-807-320

Received: 18 September 2018; Accepted: 10 November 2018; Published: 20 November 2018

Abstract: High-density ruthenium (Ru) thin films were deposited using Ru(EtCp)$_2$ (bis(ethylcyclopentadienyl)ruthenium) and oxygen by thermal atomic layer deposition (ALD) and compared to magnetron sputtered (MS) Ru coatings. The ALD Ru film growth and surface roughness show a significant temperature dependence. At temperatures below 200 °C, no deposition was observed on silicon and fused silica substrates. With increasing deposition temperature, the nucleation of Ru starts and leads eventually to fully closed, polycrystalline coatings. The formation of blisters starts at temperatures above 275 °C because of poor adhesion properties, which results in a high surface roughness. The optimum deposition temperature is 250 °C in our tool and leads to rather smooth film surfaces, with roughness values of approximately 3 nm. The ALD Ru thin films have similar morphology compared with MS coatings, e.g., hexagonal polycrystalline structure and high density. Discrepancies of the optical properties can be explained by the higher roughness of ALD films compared to MS coatings. To use ALD Ru for optical applications at short wavelengths (λ = 2–50 nm), further improvement of their film quality is required.

Keywords: atomic layer deposition; sputtering; ruthenium; thin film; optical properties; structural properties; soft X-ray; XUV

1. Introduction

Ultrathin metal films are essential for numerous applications, especially in microelectronics [1], heterogeneous catalysis [2], soft X-ray optics, and sensing. Ruthenium, as a relatively low-cost noble metal, is an attractive material when high oxidation resistance is needed [3]. Smooth and high-density Ru thin films are a preferred solution, for example, as electrodes for dynamic random access memories (DRAM) [4–7], metal-oxide-semiconductor field-effect transistors (MOSFET) [8], metal-insulator-metal capacitors [9,10], and grazing-incidence soft X-ray mirrors [11]. The atomic layer deposition (ALD) technology enables pinhole-free and conformal films with sub-nanometer thickness control. Since conventional physical vapor deposition (PVD) technologies cannot realize conformal coatings on complex shaped substrates, ALD is being considered as a promising technology for optical coatings. High-efficiency metal wire polarizers for UV spectral range have already been realized based on frequency doubling technique with iridium (Ir) coatings by ALD [12,13]. Iridium ALD coating has been also applied for Fresnel zone plates for X-ray microscopes at 1 keV synchrotron radiation [14]. Although ALD of Ru processes has been widely investigated for their electronic properties [2,6–9,15–19], their optical properties have not been analyzed yet. The large interest for Ru ALD arises since it is considered as a favorable capacitor electrode in DRAM [3,4,6,18], as gates in metal oxide semiconductor transistors [5,8], or is applied as nucleation seed layers for copper interconnect formation [19]. As an optical coating, Ru is currently realized for capping layers [3,20,21] and grazing incidence mirrors [22–25] by magnetron sputtering (MS).

This article presents the optical and structural properties of Ru ALD in the soft X-ray and XUV (extreme UV) spectral range. Furthermore, the coating properties are compared with conventional sputtered Ru films and the potential of the ALD technology for optical applications is discussed.

2. Experimental

Ru thin film deposition was performed with an Oxford Instruments OpAL open load reactor tool. In this study, $Ru(EtCp)_2$ (bis(ethylcyclopentadienyl)ruthenium, Strem Chemicals, Kehl, Germany) and O_2 were used as metalorganic precursor and co-reactant, respectively. $Ru(EtCp)_2$ as a liquid precursor has a relatively high vapor pressure of 0.24 mbar at 80 °C [26]. This temperature was applied to bubble the precursor with 150 sccm argon (Ar) gas flow into the ALD reactor. Under a working pressure of approximately 0.1 mbar, the coatings were fabricated in a thermal ALD process at substrate temperatures between 200 °C and 300 °C. Major reaction products during this ALD process are carbon dioxide (CO_2) and water (H_2O). The overall chemical reaction can be suggested as:

$$Ru(EtCp)_2 \ (g) + {}^{37}/_2 \ O_2 \ (g) \rightarrow Ru \ (s) + 9 \ H_2O \ (g) + 14 \ CO_2 \ (g) \tag{1}$$

Growth rate experiments of the Ru films were performed at a deposition temperature of 250 °C. The ALD cycle consists of four repeated steps: $Ru(EtCp)_2$ precursor pulse, precursor purge with Ar (150 sccm), co-reactant pulse with 50 sccm O_2, and a final purge with 150 sccm Ar flow. The optimized time for each step was constant at 2 s, 4 s, 3 s, and 4 s, respectively. Furthermore, a plasma enhanced ALD (PEALD) process was tested. Therefore, O_2 plasma was ignited with 100 W RF power at 100 sccm flow rate. The substrate was exposed for three seconds to the O_2 plasma instead of the thermally activated O_2 gas flow. Super-polished silicon (Si) wafers with a crystal orientation of (100) and amorphous fused silica (SiO_2) were used as conventional substrates for optical coatings.

For comparison, Ru coatings were fabricated on the DC-magnetron sputtering system NESSY 3 [27]. In this industrial system, coatings were deposited in an Ar atmosphere on Si substrates with a working pressure of 10^{-3} mbar and a source power of 500 W.

All samples produced were measured by grazing incidence X-ray reflectometry (XRR) with Cu-Kα radiation (λ = 0.154 nm) to characterize the coating properties. The XRR data were fitted with a simple single layer model (Ru on substrate, whereby the roughness of the coating is also considered) using the Leptos 7 software package (Bruker Corporation) [28]. The extracted simulation results provide information on the coating thickness, coating density, and surface roughness. The same measurement setup was used for X-ray diffraction experiments (XRD). The crystal sizes were estimated according to the Scherrer equation [29].

Furthermore, the surface was investigated with a Carl Zeiss Σigma scanning electron microscope (SEM, Carl Zeiss, Oberkochen, Germany) with a constant acceleration voltage of 10 kV and energy dispersive X-ray analysis (EDX) for chemical characterization. Surface roughness analysis was additionally carried out through atomic force microscopy (AFM, Dimension 3100 with Nanoscope IV controller, Digital Instruments, Santa Barbara, CA, USA) measurements.

The XUV reflectometry (XUVR) was carried out by the Physikalisch Technische Bundesanstalt (PTB, Bessy II, Beamline PTB-EUV, Berlin, Germany) [30] at a fixed grazing incidence angle of 10° varying the wavelength between 2 nm and 25 nm. The reflectivity curves were simulated with the

IMD-software [31] using the optical constants of Henke et al. [32]. Similar to the XRR simulations, a single-layer model has initially been applied, but had to be extended by a thin RuO_2 and C layer.

The surface composition was studied with X-ray photoelectron spectroscopy (XPS, XR 50 M X-ray source with FOCUS 500 monochromator, SPECS Surface Nano Analysis GmbH, Berlin, Germany) using an ultrahigh vacuum (UHV) surface analysis system. The photoelectrons were excited by monochromatic Al-Kα radiation (E = 1486.71 eV) under 55° angle of incidence and detected with a PHOIBOS 150 hemispherical electron analyzer (SPECS Surface Nano Analysis GmbH, Berlin, Germany).

Additionally, an Auger electron spectroscopy (AES, Varian Vacuum Division, Palo Alto, CA, USA) depth profile was performed with an Auger cylindrical mirror spectrometer. A focused 5 keV electron beam under an angle of incidence of 30° and a cylindrical mirror analyzer (CMA) were used. Sputtering was carried out with krypton (Kr) at an energy of 2 keV and a current of 10 µA.

3. Results

Atomic layer deposition of Ru has been reported using several metalorganic precursors. Hämäläinen et al. presented a review of the reported Ru ALD processes with various precursors [33]. They summarized growth rates, deposition temperatures and the evaporation temperature of different precursors relating to the corresponding co-reactants. Besides $Ru(Cp)_2$ [6,34] and $Ru(Thd)_3$ [33,35], $Ru(EtCp)_2$ [5,10,34,36] is the most commonly used precursor. Its flexible deposition properties have led to the choice of $Ru(EtCp)_2$ to start here the optical coating development. Besides thermal and plasma enhanced ALD, the precursor enables the deposition with a wide selection of co-reactants, e.g., air [6], O_2 [26,34], ozone (O_3) [10], ammonia (NH_3) [19,34], and hydrogen (H_2) [5]. The main by-products of the chosen process ($Ru(EtCp)_2$ and O_2) are water and carbon dioxide [37] (see Equation (1)) and thus it fulfills safety requirements. Furthermore, the liquid precursor $Ru(EtCp)_2$ is readily available in adequate ALD bubblers for precursor delivery.

3.1. Structural Properties

Immediately after deposition, all coatings were characterized by XRR to determine the density (ρ), roughness (σ), and film thickness (z) without major influence of contaminations.

The experimental data and the corresponding simulation are presented in Figure 1. As shown, a simple simulation model (single Ru layer on Si-substrate) describes the experimental data very well. After 1500 cycles at 250 °C, the Ru grows to a z = 57.5 nm thick film with a surface roughness of σ = 2.6 nm. At comparable temperatures of 275 °C to 300 °C, other authors reported an even higher surface roughness from 3.7 nm [34] up to 13.9 nm [10] of thermally deposited Ru. Furthermore, the Ru ALD layer exhibits a high density of ρ = 12.3 g·cm^{-3} comparable to the bulk Ru value of 12.45 g·cm^{-3} [38].

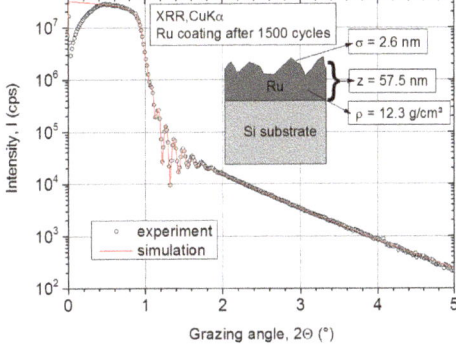

Figure 1. X-ray reflectometry (XRR) measurement on atomic layer deposition (ALD) Ru layer and corresponding simulation with a simple two-layer model (substrate + Ru thin film).

To determine the growth rate per cycle (GPC), processes were carried out with 500 cycles up to 3000 cycles and the films were thoroughly characterized. Figure 2 points out a linear thickness evolution with increasing number of ALD cycles. The GPC was determined by a linear fit to (0.047 ± 0.002) nm/cycle on a Si substrate and (0.049 ± 0.002) nm/cycle on fused silica, respectively. Compared with other ALD processes using Ru(EtCp)$_2$ as a precursor, the determined GPCs are of the same magnitude. Wojcik et al. reported a growth rate of 0.037 nm/cycle with a Ru(EtCp)$_2$ PEALD process [39] and Park et al. 0.075 nm/cycle with a thermal process [34]. A higher growth rate of 0.12 nm/cycle was achieved with O$_3$ as co-reactant [10].

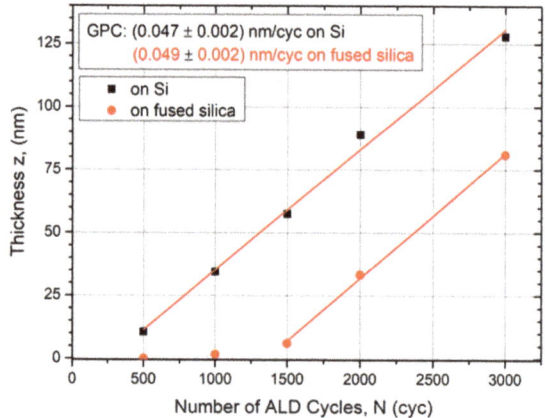

Figure 2. Evolution of Ru film thickness determined by XRR measurements as function of the number of ALD cycles on different substrates. After a nucleation delay, the growth rate per cycle (GPC) becomes constant for both substrate materials.

Figure 2 indicates a nucleation delay since the linear fit does not cross the origin until the growth rate becomes constant. This delay lasts longer for amorphous fused silica substrates compared with Si substrates. Following the linear fit in Figure 2, the linear growth regime starts after approximately 250 cycles on Si and after 1350 cycles on SiO$_2$, respectively. An ALD review by George indicates that ALD metals prefer to form clusters on oxide surfaces [40]. Depending on the substrate material, it takes a certain amount of ALD cycles until the first metal layer is fully closed and the growth becomes linear per cycle. Hämäläinen et al. reported that this nucleation delay for Ru can last up to hundreds of cycles [33], as our observation also indicates. The initial density of nucleation sites plays a critical role in the formation of Ru thin films and is higher on Si than on SiO$_2$. On noble metal substrates, such as platinum (Pt) and palladium (Pd), no nucleation delay was found by Lu et al. [5].

Nevertheless, all ALD coatings show a high density of (12.3 ± 0.1) g·cm^{-3} by XRR independent of their thickness. Repeated XRR measurements show identical curves three months past deposition. Thus, high coating stability is assumed due to identical XRR results. High-density coatings are important for the optical properties in the soft X-ray and XUV spectral range [41]. Kim et al. report a density of ρ = 11.9 g·cm^{-3} by XRR investigations [10] although they used O$_3$ as a more reactive co-reactant. The highest Ru coating density of 12.7 g·cm^{-3}, achieved with an O$_2$ based ALD process, was reported by Manke et al. [42] at temperatures of T_{dep} = 450 °C. In addition to the density, a low surface roughness is also essential for high reflective properties in optical applications. We have determined a strong influence of the deposition temperature T_{dep} on the film growth and thus the surface roughness.

Figure 3 presents film surface morphologies after 500 cycles deposited at different temperatures. No deposition was observed at temperatures below T_{dep} < 200 °C. The deposition started at T_{dep} = 230 °C. However, the nucleation delay was so high that even after 500 cycles the layer was not fully covering the substrate and Ru nanoparticles were clearly visible. The black background corresponds to the Si substrate surface. The small white areas on top are Ru nuclei with a diameter of approximately 6–20 nm. The growth rate rises with an increasing temperature of T_{dep} = 250 °C. At this point, the complete surface is covered with Ru crystals and results in a film thickness of z = 17 nm. Furthermore, the estimated crystal size is of a similar magnitude (20–30 nm). At a higher deposition temperature of T_{dep} = 300 °C, the lateral crystal size grows further to 25–40 nm. Besides a larger film thickness (z = 35 nm by XRR) caused by a higher growth rate, a possible reason for larger grains is the higher substrate temperature itself. This increases the mobility of the Ru atoms and thus they can reach places at a favorable energetic state, e.g., the crystal lattice.

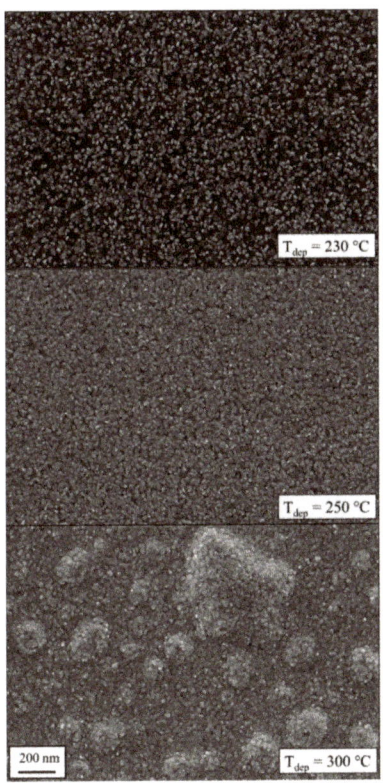

Figure 3. Scanning electron microscopy (SEM) images of Ru layers at different deposition temperatures after 500 ALD cycles. While no deposition was observed below T_{dep} = 200 °C, blistering occurred at higher temperatures. A temperature optimum was found at 250 °C.

Furthermore, the formation of blisters at deposition temperature of 300 °C can be seen in Figure 3. These large blisters consist of air pockets. This phenomenon was also reported by Kim et al. for a Ru(EtCp)$_2$–O$_2$ process [10]. For Ru deposition at T_{dep} = 275 °C, they reported a formation of blisters that led to a significant surface roughness of approximately 14 nm [10]. Other O$_2$ based metal ALD, e.g., Ir [43], show also the effect of blistering, whereby improving the process parameters (e.g., long purge time) significantly minimized the appearance of defects. Gadkari et al. described that the blistering can be caused by the combination of stress and a weak film-substrate adhesion [44]. Nevertheless, we also detected that the film adhesion became worse and the coating could easily be scratched. Thus, these coatings deposited at high temperatures are not suitable for optical applications.

As shown in Figure 3, all coatings demonstrate a crystalline growth. Therefore, additional XRD-investigations have been carried out to study the crystallinity of the Ru ALD coatings in detail. Figure 4 shows the diffraction pattern of a 57 nm thick coating on a Si substrate deposited at the optimized temperature of 250 °C. For comparison, an MS coating with equivalent thickness is additionally presented.

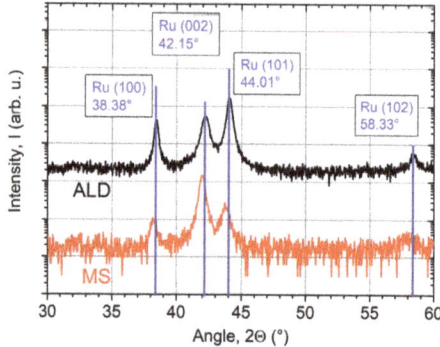

Figure 4. X-ray diffraction (XRD) pattern of ALD (z = 57 nm) and magnetron sputtered (MS) (z = 50 nm) Ru coatings on Si substrates with corresponding phase analysis. Diffraction peaks of the MS coating are slightly shifted to smaller angles indicating film stress.

Both coatings consist of hexagonal polycrystals [45]. Three peaks can be clearly identified at the angles of 38°, 42°, and 44°. The fourth peak at 58° is only indicated because of its low intensity. The blue solid lines show the intensity distribution of each peak for a randomly oriented powder sample. This indicates that both coatings are randomly oriented with the Ru polycrystalline structure oriented in the [100], [002] and [101] orientations in the growth direction. Further, the peaks of the sputtered coating are slightly shifted to smaller angles. These findings suggest that the coating is under compressive stress. Alagoz et al. report on high compressive stresses of several GPa in Ru films by MS [46]. The ALD sample matches the diffraction database values perfectly. Therefore, low film stress is expected. The film stress, reported by Kim et al. based on a Ru(EtCp)$_2$ and O$_2$ process, was also low with a value of 88 MPa [10].

Figure 5 summarizes the evolution of grain size and roughness with increasing film thickness. The grain size was estimated by the Scherrer equation based on the (101) peak with maximum intensity. When the film thickness increases, the grain size rises as well. Nevertheless, for thin films, the grain growth evolves further into a saturation of approximately Λ = 30 nm. The MS Ru coatings have a comparable grain size (Λ = 21 nm, z = 50 nm).

The estimated grain sizes by the Scherrer equation from XRD are commonly smaller than from SEM images because only coherently scattering areas contribute to the signal. Other effects, e.g., mechanical stress, can also lead to a broadening of Bragg peaks and thus apparently smaller grains. Hence, the Scherrer values are considered as a lower limit value.

In relation, the surface roughness σ of all ALD coatings with thicknesses between 10 and 130 nm show similar results between 2 and 4 nm (AFM measurement). Interestingly, the thinnest layers demonstrate high roughness values that are as high as for thick layers. Due to the nucleation process, we assume the appearance of a Ru film which is not a fully closed layer with protruding crystals. This effect could cause the high roughness at the early observation stage. As the film thickness increases, the layer closes and smooths out the surface to a roughness of only 2.6 nm for a 57 nm thick coating. When the film thickness further increases, the surface roughness rises as well, but no blisters have been observed. This can be explained by larger Ru crystals. Due to the fact that the crystal size growth saturates, we also expect a saturated roughness of coatings with thicknesses larger than z > 150 nm. The MS Ru coating is much smoother, although the crystal size has comparable dimensions (σ = 0.5 nm,

Λ = 21 nm) compared to the ALD sample (σ = 2.6 nm, Λ = 26 nm). All investigated coating properties for a 50 nm thick film are summarized and compared with an MS coating in Table 1.

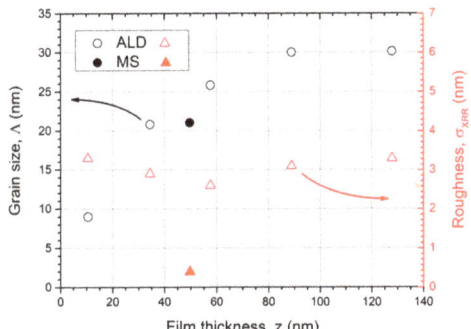

Figure 5. The estimated grain size by the Scherrer equation (left, ○) increases with coating thickness and saturates at Λ = 30 nm. The roughness evolution with increasing Ru film thickness is additionally shown (right, △). The filled symbols correspond to the MS coating.

Table 1. Comparison of coating properties of a 57.5 nm thick ALD and a 50 nm thick MS Ru coating.

Property	Atomic Layer Deposition (ALD)	Magnetron Sputtering (MS)
Roughness, σ_{XRR} (nm)	2.6 ± 0.5	0.4 ± 0.1
Density, ρ_{XRR} (g·cm^{-3})	12.3 ± 0.1	12.4 ± 0.1
Structure	polycrystal hexagonal	polycrystal hexagonal
Grainsize, Λ (nm)	26	21

3.2. Optical Properties

The measurement of the reflectance in the XUV spectral range requires specialized beamline equipment, which is only available in a few research centers. Hence, the XUV reflectance measurement could be carried out only 5 months after deposition of the ALD samples. Degradation of the Ru ALD samples with time cannot be fully excluded. The thickness of the Ru thin films for XUV applications at a grazing angle of incidence should exceed tens of nanometers to obtain high reflectance values. Further, for numerous reasons, even film thicknesses >200 nm are required to ensure the stability of the mirror optics at the beamline to minimize damage due to the radiation. In Figure 6, the reflection data determined for an ALD sample with a layer thickness of z = 35 nm is compared to a sputtered Ru film (z = 50 nm). A lower reflection appears for the fixed angle of incidence Θ = 10° in the spectral range between λ = 2 nm and 25 nm (Figure 6) for the ALD coating as compared to the MS sample. In addition to the normalized integral reflection, the critical angle for the ALD film is significantly smaller (see Table 2).

Although a simple single-layer model leads to a good agreement with the experiment for the XRR simulations, the model had to be improved by two additional layers besides surface roughness to fit the XUV reflectance data. First, a ruthenium(IV) oxide layer (RuO_2: ρ = 7.0 g·cm^{-3}) on the Ru surface was assumed, and second carbon residuals (C: ρ = 2.0 g·cm^{-3}). These thin surface layers could not be detected by XRR due to the high surface roughness. The characteristic oscillation is suppressed in XRR even at small angles of incidence. Hence, the features of the thin surface layer, which typically occur at large angles, are not visible.

With the extended model, the XUVR measurement data could be fitted very well. The fit shows that the surface of the ALD coating is significantly rougher (σ_{ALD} = (4.5 ± 0.6) nm) than that of the sputtered film (σ_{MS} = (0.7 ± 0.3) nm). These results are in qualitative agreement with the XRR study of the roughness. The surface roughness is a critical factor affecting the reflectivity at XUV wavelength range because the microstructural relevant size of the surface is close to the wavelength of light. In terms of reflectivity, the higher roughness of the ALD coating leads to a reduction of nearly 20% compared to the MS coating. In the XUVR simulations, the thickness of the surface oxide on the ALD film is larger (d_{oxide} = 2.5 nm) compared to the sputtered layer (d_{oxide} = 1.0 nm). Due to the increased surface roughness and grainy topography of the ALD sample, it is assumed as the cause of the increased oxide thickness. Furthermore, a 1.6 nm thick C layer was fitted on the RuO_2 layer. This third layer was not presumed for the simulation of the sputtered sample. The C residuals on the surface of the ALD layer could be attributed to reaction products during the ALD process because C is a component of the precursor $Ru(EtCp)_2$. Additionally, the C surface contamination due to increased roughness and long storage time is probable.

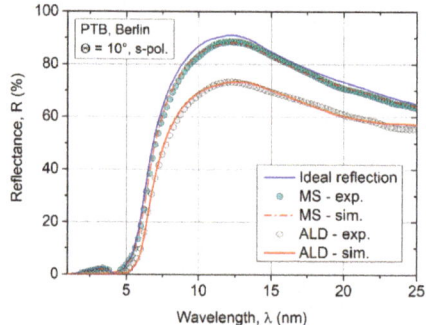

Figure 6. Comparison of experimental reflection data of sputtered and ALD Ru films in the soft X-ray and extreme UV (XUV) spectral range. Shown in red is the fit with the simulation model Ru + RuO_2 + C. In blue, the reflectivity without a surface oxide and ideal smooth interfaces is shown.

Table 2. Comparison of the optical properties and simulation results for a 35 nm thick ALD and a 50 nm thick sputtered Ru layer from XUV reflectometry (XUVR) measurements.

Property	ALD	MS
Int. reflection [2.5°–20.0°], R_{exp} (°)	12.1	14.9
Normalized int. reflection, R_{exp}/R_{ideal}	0.79	0.97
Critical angle, Θ_{crit} (°)	18.4	25.3
Surface roughness, σ (nm)	4.6 ± 0.6	0.7 ± 0.6
Thickness of oxide layer, d_{oxide} (nm)	2.5 ± 0.5	1.0 ± 0.5
Carbon residuals, d_C (nm)	1.5 ± 0.5	–

If the influences of the surface roughness σ and the layer thicknesses (RuO_2, C) for the ALD coating are considered separately, σ describes three-quarters of the reflection losses with respect to the normalized integral reflection R_{exp}. If the reflection curve is simulated with a RuO_2 and C layer, it leads to a quarter of the reflection losses, neglecting the surface roughness (σ = 0 nm). This observation shows that the roughness of the Ru ALD film is the critical reflection-reducing effect. The simulation results with the optical properties are summarized in Table 2.

3.3. Chemical Analysis

To verify the XUVR simulation model, XPS and an AES depth profile were performed on a z = 57 nm thick ALD sample. With both methods, only Ru and O were detected. The XPS survey spectra with indicated features are shown in Figure 7. Oxygen is mainly present on the surface because the Ru features decrease stronger than the O features with increasing angle of photoelectron emission (increased surface sensitivity). The Ru $3d_{3/2}$ state overlaps the C 1s state at $E_B \approx$ 285 eV (see Figure 7 inset). Unfortunately, a determination of the amount of adventitious C on the surface is not possible because the C 1s peak is obstructed by the Ru $3d_{3/2}$ feature.

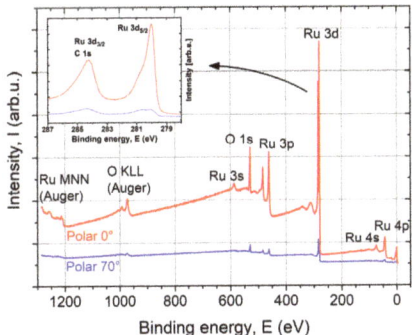

Figure 7. X-ray photoelectron spectroscopy (XPS) survey spectra of ALD Ru (z = 57 nm) coating on a Si substrate at normal emission (0°) and a polar angle of 70° with indicated features. Only Ru and O were detected, with O mainly being present on the layer surface. The inset shows a zoom-in into the Ru 3d state region.

The Ru $3d_{5/2}$ peak appears asymmetric at normal emission (0°) and has clearly two components at 70°. This originates from metallic Ru and oxidized Ru. Based on the binding energies of different Ru oxides, the surface oxide can be confirmed as RuO_2 corresponding to the shoulder around 280.8 eV (see Table 3). The thickness of the surface RuO_2 layer can be estimated between 1 nm and 2 nm, which is in agreement with the XUVR simulation.

Table 3. Binding energies of Ru $3d_{5/2}$ state for metallic Ru and different Ru oxides [46].

Compound	Ru $3d_{5/2}$ Binding Energy [eV]
Ru	≈280.0
RuO_2	≈280.8
RuO_3	≈282.5
RuO_4	≈283.2

Figure 8 presents an atomic ratios depth profile for Ru and O using AES. On the surface, that means without sputtering, more than 60 at % O is situated which fits to a thin RuO_2 on the surface. The O content decreases exponentially with increasing sputter depth and remains at 0.8% in the bulk material. Probably due to the high surface roughness, no ideal layered structure of RuO_2 and Ru occurs. Carbon was not measurable, because the Ru signal overlaps spectrally the C signal again.

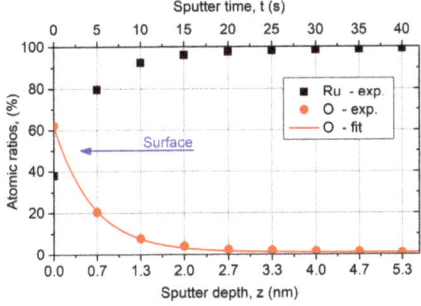

Figure 8. Atomic ratios depth profile for Ru and O of ALD Ru (z = 57 nm) coating on a Si substrate using Auger electron spectroscopy (AES). The composition on the surface fits RuO_2, whereby the O content decreases fast with increasing sputter depth and remains at 0.8% in the bulk material. The sputter depth is a rough approximation, whereby the sputter rate is roughly 8 nm/min.

4. Discussion

Ruthenium thin films deposited by ALD have been thoroughly characterized. The typical ALD temperature window to obtain a constant growth rate for the specific process is in the range of 230 to 300 °C. However, the nucleation and film properties significantly depend on the deposition temperature.

The results obtained in this study indicate that further optimization of ALD Ru coatings for optical applications is essential to leverage on the ALD capability of conformal coatings on nanostructured substrates or on complex shaped optics. The comparison between the ALD and MS technologies has pointed out several advantages of the MS technology. As shown, a sputtered Ru coating resulted in a highly dense and smooth film. The reflection measurements and simulations (Figure 6) indicate that the smooth interfaces are essential for high reflection properties at short wavelengths. The reflectance of the sputtered coatings is approximately 20% higher than the ALD coatings at the peak wavelength of 12 nm. However, there are also disadvantages for MS compared to ALD. The XRD pattern suggests higher film stress in MS films. Stress can lead to deviations of manufacturing tolerances or delamination. Alagoz et al. reported that highly dense Ru films could not be grown to a thickness of higher than 85 nm by magnetron sputtering because they started to peel off beyond this thickness [47]. In this study, Ru coating with a film thickness of 120 nm has been achieved by ALD although it is a relatively slow deposition process. In general, the crystal size and roughness continuously grow with increasing film thickness in MS coatings [48], whereas in ALD the crystal size and surface roughness do not alter with increasing coating thickness. The roughness evolution of MS films, caused by grain growth, increases significantly in thicker coatings in contrast to ALD. The technology of ALD has further advantages compared with conventional MS. A conformal deposition is possible over large areas and even on complex-shaped substrates [49]. Furthermore, there is the possibility to switch easily between Ru and RuO_2 coatings while increasing the O_2 pressure and pulse lengths [18].

Concerning the optical properties, the higher surface roughness of ALD coatings mainly reduces the reflection at short wavelengths. We tried to reduce the crystal growth to smaller grains and thus the roughness evolution. While Park et al. reported smoother surfaces of $\sigma = 0.9$ nm with an NH_3 based PEALD process [34], there was no improvement observed using O_2 plasma. No film growth was observed by PEALD with O_2 plasma. The O_2 plasma was strong enough to instantaneously remove the grown Ru layer. Similar etching phenomena were reported by Belau et al. [20]. Even by changing the deposition temperature, a further improvement of the surface roughness was not possible, and a deposition with NH_3 is currently not available in our laboratories.

As potential option to improve the nucleation seed density and thus the fast formation of the first fully closed metal layer, a metal-based seed layer can be implemented as Lu et al. reported that no nucleation delay was found on Pt and Pd layers [5]. Furthermore, Kim et al. showed a nucleation improvement and a surface roughness reduction by O_3 supply [10]. An alternative chemistry of precursor and co-reactant is also possible. There are several other precursors and processes that achieved smoother Ru ALD films, e.g., $Ru(MePy)_2$: $\sigma = 0.2$ nm [50], RuEtPy: $\sigma = 0.4$ nm [51] and RuO_4 with H_2: $\sigma = 0.3$ nm. However, with regard to other necessary requirements for optical applications, e.g., high purity and high density, these processes need to be evaluated in detail.

The chemical analysis of the ALD Ru surface has confirmed the presence of a thin oxide layer at the surface of the coating; however, not within the film. This ultra-thin surface layer also significantly affects the XUV reflection of the ALD Ru, even though not as much as the high surface roughness. The oxide layer might be related to the chemical process which involves an oxidizing step to remove the organic ligands of the $Ru(EtCp)_2$ precursor. In the O_2 pulse step, the Ru would be also partially oxidized. In the following metalorganic pulse, the oxide is probably decomposed and the precursor ligands serve as a reducing agent. Hence, little oxygen contamination is being detected in the film. In contrast, at the surface, the reaction terminates with the O_2 pulse and remaining RuO_2 is not further reduced. This further motivates the development of oxygen free ALD processes of metals.

5. Conclusions

Ruthenium thin films grown by ALD have been evaluated for optical applications. High-density Ru films have successfully been deposited on Si and fused silica substrates. Temperatures above 230 °C are required for a film formation using Ru(EtCp)$_2$ and O$_2$. Higher deposition temperatures above 275 °C have led to blisters that increased the surface roughness significantly and reduced the adhesion properties. With an optimized deposition temperature of 250 °C, we have explored the film growth and the resulting optical properties. The polycrystalline growth and the corresponding evolution of the surface roughness have led to major reflection losses at short wavelengths. Sputtered Ru coatings show similar density but are much smoother than ALD coatings. The ALD samples show a thin (<2 nm) RuO$_2$ surface layer and 0.8 at % residual O in bulk material. Carbon impurities were not measurable. Further experimental development is required to leverage on the major benefit of ALD to realize conformal coatings on complex-shaped substrates towards coatings with high reflectance and stability for optical applications.

Author Contributions: Conceptualization, R.M., L.G., and A.S.; ALD Depositions, L.G.; SEM Characterization, S.W.; AFM Measurements, V.B.; XRR and XUV Analysis, R.M.; XPS and AES Analysis, F.O. and T.F.; Writing–Original Draft Preparation, R.M. and P.S.; Writing–Review and Editing, P.S., L.G., T.F, and A.S.; Supervision, A.S.; Project Administration, A.S., N.K. and A.T.; Funding Acquisition, A.S., N.K. and A.T.

Funding: This research was funded by the Fraunhofer Society Attract Project (No. 066-601020), the Fraunhofer IOF, Center of Excellence in Photonics and the Deutsche Forschungsgemeinschaft (DFG) Emmy-Noether-Project (No. SZ235/1-1).

Acknowledgments: The authors thank David Kästner for the technical support and Philipp Naujok for helpful discussions. We also acknowledge constructive comments of reviewers to improve the article.

Conflicts of Interest: The authors declare no conflict of interest. The funders had no role in the design of the study; in the collection, analyses, or interpretation of data; in the writing of the manuscript, or in the decision to publish the results.

References

1. Yeo, S.; Choi, S.-H.; Park, J.-Y.; Kim, S.-H.; Cheon, T.; Lim, B.-Y.; Kim, S. Atomic layer deposition of ruthenium (Ru) thin films using ethylbenzen-cyclohexadiene Ru (0) as a seed layer for copper metallization. *Thin Solid Films* **2013**, *546*, 2–8. [CrossRef]
2. Minjauw, M.M.; Dendooven, J.; Capon, B.; Schaekers, M.; Detavernier, C. Near room temperature plasma enhanced atomic layer deposition of ruthenium using the RuO$_4$-precursor and H$_2$-plasma. *J. Mater. Chem. C* **2015**, *3*, 4848–4851. [CrossRef]
3. Bajt, S.; Dai, Z.R.; Nelson, E.J.; Wall, M.A.; Alameda, J.; Nguyen, N.; Baker, S.; Robinson, J.C.; Taylor, J.S.; Clift, M.; et al. Oxidation resistance of Ru-capped EUV multilayers. *Proc. SPIE* **2005**, *5751*, 118–127. [CrossRef]
4. Aoyama, T.; Eguchi, K. Ruthenium films prepared by liquid source chemical vapor deposition using bis-(ethylcyclopentadienyl)ruthenium. *Jpn. J. Appl. Phys.* **1999**, *38*, L1134. [CrossRef]
5. Lu, J.; Elam, J.W. Low temperature ABC-type Ru atomic layer deposition through consecutive dissociative chemisorption, combustion, and reduction steps. *Chem. Mater.* **2015**, *27*, 4950–4956. [CrossRef]
6. Aaltonen, T.; Alén, P.; Ritala, M.; Leskelä, M. Ruthenium thin films grown by atomic layer deposition. *Chem. Vap. Depos.* **2003**, *9*, 45–49. [CrossRef]
7. Lee, J.; Song, Y.W.; Lee, K.; Lee, Y.; Jang, H.K. Atomic layer deposition of Ru by using a new Ru-precursor. *ECS Trans.* **2006**, *2*, 1–11. [CrossRef]
8. Misra, V.; Lucovsky, G.; Parsons, G. Issues in high-K gate stack interfaces. *MRS Bull.* **2002**, *27*, 212–216. [CrossRef]
9. Leick-Marius, N. Atomic layer deposition of ruthenium films: Properties and surface reactions. Ph.D. Thesis, Technische Universiteit Eindhoven, Eindhoven, The Netherlands, 2014.
10. Kim, J.-Y.; Kil, D.-S.; Kim, J.-H.; Kwon, S.-H.; Ahn, J.-H.; Roh, J.-S.; Park, S.-K. Ru films from bis(ethylcyclopentadienyl)ruthenium using ozone as a reactant by atomic layer deposition for capacitor electrodes. *J. Electrochem. Soc.* **2012**, *159*, H560–H564. [CrossRef]

11. Marczuk, P.; Egle, W. Source collection optics for EUV lithography. In *Advances in Mirror Technology for X-ray, EUV Lithography, Laser, and Other Applications II*; Khounsary, A.M., Dinger, U., Ota, K., Eds.; SPIE Optical Engineering: Bellingham, WA, USA, 2004; Volume 5533, pp. 145–157.
12. Weber, T.; Käsebier, T.; Szeghalmi, A.; Knez, M.; Kley, E.-B.; Tünnermann, A. Iridium wire grid polarizer fabricated using atomic layer deposition. *Nanoscale Res. Lett.* **2011**, *6*, 558. [CrossRef] [PubMed]
13. Weber, T.; Käsebier, T.; Szeghalmi, A.; Knez, M.; Kley, E.-B.; Tünnermann, A. High aspect ratio deep UV wire grid polarizer fabricated by double patterning. *Microelectron. Eng.* **2012**, *98*, 433–435. [CrossRef]
14. Jefimovs, K.; Vila-Comamala, J.; Pilvi, T.; Raabe, J.; Ritala, M.; David, C. Zone-doubling technique to produce ultrahigh-resolution X-ray optics. *Phys. Rev. Lett.* **2007**, *99*, 264801. [CrossRef] [PubMed]
15. Shibutami, T.; Kawano, K.; Oshima, N.; Yokoyama, S.; Funakubo, H. A novel ruthenium precursor for MOCVD without seed ruthenium layer. *MRS Proc.* **2002**, *748*, U12.7. [CrossRef]
16. Wang, H.; Gordon, R.G.; Alvis, R.; Ulfig, R.M. Atomic layer deposition of ruthenium thin films from an amidinate precursor. *Chem. Vap. Depos.* **2009**, *15*, 312–319. [CrossRef]
17. Minjauw, M.M.; Dendooven, J.; Capon, B.; Schaekers, M.; Detavernier, C. Atomic layer deposition of ruthenium at 100 °C using the RuO_4-precursor and H_2. *J. Mater. Chem. C* **2015**, *3*, 132–137. [CrossRef]
18. Methaapanon, R.; Geyer, S.M.; Lee, H.-B.-R.; Bent, S.F. The low temperature atomic layer deposition of ruthenium and the effect of oxygen exposure. *J. Mater. Chem.* **2012**, *22*, 25154–25160. [CrossRef]
19. Kwon, O.-K.; Kwon, S.-H.; Park, H.-S.; Kang, S.-W. PEALD of a Ru adhesion layer for Cu interconnects. *J. Electrochem. Soc.* **2004**, *151*, C753–C756. [CrossRef]
20. Belau, L.; Park, J.Y.; Liang, T.; Somorjai, G.A. The effects of oxygen plasma on the chemical composition and morphology of the Ru capping layer of the extreme ultraviolet mask blanks. *J. Vac. Sci. Technol. B* **2008**, *26*, 2225–2229. [CrossRef]
21. Hill, S.B.; Ermanoski, I.; Tarrio, C.; Lucatorto, T.B.; Madey, T.E.; Bajt, S.; Fang, M.; Chandhok, M. Critical parameters influencing the EUV-induced damage of Ru-capped multilayer mirrors. *Proc. SPIE* **2007**, *6517*. [CrossRef]
22. Nieto, M.; Allain, J.-P.; Titov, V.; Hendricks, M.R.; Hassanein, A.; Rokusek, D.; Chrobak, C.; Tarrio, C.; Barad, Y.; Grantham, S.; et al. Effect of xenon bombardment on ruthenium-coated grazing incidence collector mirror lifetime for extreme ultraviolet lithography. *J. Appl. Phys.* **2006**, *100*, 053510. [CrossRef]
23. Shin, H.; Sporre, J.R.; Raju, R.; Ruzic, D.N. Reflectivity degradation of grazing-incident EUV mirrors by EUV exposure and carbon contamination. *Microelectron. Eng.* **2009**, *86*, 99–105. [CrossRef]
24. Zocchi, F.E.; Benedetti, E. Optical designs of grazing incidence collector for extreme ultraviolet lithography. *J. Micro/Nanolithogr. MEMS MOEMS* **2007**, *6*, 043002. [CrossRef]
25. Sweatt, W.C.; Kubiak, G.D. Condenser for Extreme-UV Lithography with Discharge Source. U.S. Patent 6,285,737 B1, 4 September 2001.
26. Kwon, O.-K.; Kim, J.-H.; Park, H.-S.; Kang, S.-W. Atomic layer deposition of ruthenium thin films for copper glue layer. *J. Electrochem. Soc.* **2004**, *151*, G109–G112. [CrossRef]
27. Sergey, Y. Multilayer Coatings for EUV/Soft X-ray Mirrors. In *Optical Interference Coatings*; Kaiser, N., Pulker, H.K., Eds.; Springer: Berlin, Germany, 2003; p. 299.
28. LEPTOS 7.8 (2014). Available online: https://www.bruker.com/products/x-ray-diffraction-and-elemental-analysis/x-ray-diffraction/xrd-software/leptos/leptos-r.html (accessed on 10 November 2018).
29. Scherrer, P. Bestimmung der Größe und der inneren Struktur von Kolloidteilchen mittels Röntgenstrahlen. In *Kolloidchemie Ein Lehrbuch. Chemische Technologie in Einzeldarstellungen*; Springer: Berlin/Heidelberg, Germany, 1912; pp. 387–409.
30. Scholze, F.; Tümmler, J.; Ulm, G. High-accuracy radiometry in the EUV range at the PTB soft x-ray beamline. *Metrologia* **2003**, *40*, S224. [CrossRef]
31. Windt, D.L. IMD-Software for modeling the optical properties of multilayer films. *Comput. Phys.* **1998**, *12*, 360–370. [CrossRef]
32. Henke, B.L.; Gullikson, E.M.; Davis, J.C. X-ray interactions: Photoabsorption, scattering, transmission, and reflection at E = 50–30,000 eV, Z = 1–92. *Atomic Data Nucl. Data Tables* **1993**, *54*, 181–342. [CrossRef]
33. Hämäläinen, J.; Ritala, M.; Leskelä, M. Atomic layer deposition of noble metals and their oxides. *Chem. Mater.* **2014**, *26*, 786–801. [CrossRef]

34. Park, S.-J.; Kim, W.-H.; Lee, H.-B.-R.; Maeng, W.J.; Kim, H. Thermal and plasma enhanced atomic layer deposition ruthenium and electrical characterization as a metal electrode. *Microelectron. Eng.* **2008**, *85*, 39–44. [CrossRef]
35. Aaltonen, T.; Ritala, M.; Arstila, K.; Keinonen, J.; Leskelä, M. Atomic layer deposition of ruthenium thin films from Ru(thd)$_3$ and oxygen. *Chem. Vap. Depos.* **2004**, *10*, 215–219. [CrossRef]
36. Kukli, K.; Kemell, M.; Puukilainen, E.; Aarik, J.; Aidla, A.; Sajavaara, T.; Laitinen, M.; Tallarida, M.; Sundqvist, J.; Ritala, M.; et al. Atomic layer deposition of ruthenium films from (ethylcyclopentadienyl) (pyrrolyl)ruthenium and oxygen. *J. Electrochem. Soc.* **2011**, *158*, D158. [CrossRef]
37. Aaltonen, T.; Rahtu, A.; Ritala, M.; Leskelä, M. Reaction mechanism studies on atomic layer deposition of ruthenium and platinum. *Electrochem. Solid State Lett.* **2003**, *6*, C130. [CrossRef]
38. Perry, D.L. *Handbook of Inorganic Compounds*; CRC Press: Boca Raton, FL, USA, 2011.
39. Wojcik, H.; Junige, M.; Bartha, W.; Albert, M.; Neumann, V.; Merkel, U.; Peeva, A.; Gluch, J.; Menzel, S.; Munnik, F.; et al. Physical characterization of PECVD and PEALD Ru(-C) films and comparison with PVD ruthenium film properties. *J. Electrochem. Soc.* **2012**, *159*, H166–H176. [CrossRef]
40. George, S.M. Atomic layer deposition: An overview. *Chem. Rev.* **2010**, *110*, 111–131. [CrossRef] [PubMed]
41. Spiller, E.A. *Soft X-ray Optics*; SPIE Optical Engineering: Bellingham, WA, USA, 1994; p. 235.
42. Manke, C.; Miedl, S.; Boissiere, O.; Baumann, P.K.; Lindner, J.; Schumacher, M.; Brodyanski, A.; Scheib, M. Atomic vapor deposition of Ru and RuO$_2$ thin film layers for electrode applications. *Microelectron. Eng.* **2005**, *82*, 242–247. [CrossRef]
43. Genevée, P.; Ahiavi, E.; Janunts, N.; Pertsch, T.; Oliva, M.; Kley, E.-B.; Szeghalmi, A. Blistering during the atomic layer deposition of iridium. *J. Vac. Sci. Technol. A* **2016**, *34*, 01A113. [CrossRef]
44. Gadkari, P.R.; Warren, A.P.; Todi, R.M.; Petrova, R.V.; Coffey, K.R. Comparison of the agglomeration behavior of thin metallic films on SiO$_2$. *J. Vac. Sci. Technol. A* **2005**, *23*, 1152–1161. [CrossRef]
45. *Powder Diffraction File*; Pattern 06-0663; ICDD: Newtown Square, PA, USA, 1997.
46. Moulder, J.F.; Stickle, W.F.; Sobol, P.E.; Bomben, K.D. *Handbook of X-ray Photoelectron Spectroscopy*; Physical Electronics, Inc.: Eden Prairie, MN, USA, 2009; pp. 114–115.
47. Alagoz, A.S.; Kamminga, J.-D.; Grachev, S.Y.; Lu, T.-M.; Karabacak, T. Residual stress reduction in sputter deposited thin films by density modulation. *MRS Proc.* **2009**, *1224*, 1224-FF05-22. [CrossRef]
48. Petrov, I.; Barna, P.B.; Hultman, L.; Greene, J.E. Microstructural evolution during film growth. *J. Vac. Sci. Technol. A* **2003**, *21*, S117–S128. [CrossRef]
49. Pfeiffer, K.; Schulz, U.; Tünnermann, A.; Szeghalmi, A. Antireflection Coatings for Strongly Curved Glass Lenses by Atomic Layer Deposition. *Coatings* **2017**, *7*, 118. [CrossRef]
50. Kukli, K.; Aarik, J.; Aidla, A.; Jõgi, I.; Arroval, T.; Lu, J.; Sajavaara, T.; Laitinen, M.; Kiisler, A.-A.; Ritala, M.; et al. Atomic layer deposition of Ru films from bis(2,5-dimethylpyrrolyl)ruthenium and oxygen. *Thin Solid Films* **2012**, *520*, 2756–2763. [CrossRef]
51. Geidel, M.; Junige, M.; Albert, M.; Bartha, J.W. In-situ analysis on the initial growth of ultra-thin ruthenium films with atomic layer deposition. *Microelectron. Eng.* **2013**, *107*, 151–155. [CrossRef]

© 2018 by the authors. Licensee MDPI, Basel, Switzerland. This article is an open access article distributed under the terms and conditions of the Creative Commons Attribution (CC BY) license (http://creativecommons.org/licenses/by/4.0/).

Article

Structural and Optical Properties of Luminescent Copper(I) Chloride Thin Films Deposited by Sequentially Pulsed Chemical Vapour Deposition

Richard Krumpolec [1], Tomáš Homola [1], David C. Cameron [1,*], Josef Humlíček [2], Ondřej Caha [2], Karla Kuldová [3], Raul Zazpe [4,5], Jan Přikryl [4] and Jan M. Macak [4,5]

[1] R & D Center for Low-Cost Plasma and Nanotechnology Surface Modifications (CEPLANT), Department of Physical Electronics, Faculty of Science, Masaryk University, Kotlářská 267/2, 611 37 Brno, Czech Republic; 235947@mail.muni.cz (R.K.); homola.tomas@gmail.com (T.H.)
[2] Department of Condensed Matter Physics, Masaryk University, Kotlářská 267/2, 611 37 Brno, Czech Republic; humlicek@sci.muni.cz (J.H.); caha@monoceros.physics.muni.cz (O.C.)
[3] Institute of Physics of the Czech Academy of Sciences, v.v.i., Cukrovarnická 10, 162 00 Prague 6, Czech Republic; kuldova@fzu.cz
[4] Centre of Materials and Nanotechnologies, Faculty of Chemical Technology, University of Pardubice, Nám. Čs. Legií 565, 530 02 Pardubice, Czech Republic; Raul.Zazpe@upce.cz (R.Z.); Jan.Prikryl@upce.cz (J.P.); jan.macak@upce.cz (J.M.M.)
[5] Central European Institute of Technology, Brno University of Technology, Purkyňova 123, 612 00 Brno, Czech Republic
* Correspondence: dccameron@mail.muni.cz; Tel.: +420-549-49-1411

Received: 31 August 2018; Accepted: 15 October 2018; Published: 18 October 2018

Abstract: Sequentially pulsed chemical vapour deposition was used to successfully deposit thin nanocrystalline films of copper(I) chloride using an atomic layer deposition system in order to investigate their application to UV optoelectronics. The films were deposited at 125 °C using [Bis(trimethylsilyl)acetylene](hexafluoroacetylacetonato)copper(I) as a Cu precursor and pyridine hydrochloride as a new Cl precursor. The films were analysed by XRD, X-ray photoelectron spectroscopy (XPS), SEM, photoluminescence, and spectroscopic reflectance. Capping layers of aluminium oxide were deposited in situ by ALD (atomic layer deposition) to avoid environmental degradation. The film adopted a polycrystalline zinc blende-structure. The main contaminants were found to be organic materials from the precursor. Photoluminescence showed the characteristic free and bound exciton emissions from CuCl and the characteristic exciton absorption peaks could also be detected by reflectance measurements.

Keywords: vapour deposition; copper chloride; characterization; optical properties; XPS; crystal structure

1. Introduction

Copper(I) chloride (CuCl) is a wide-bandgap, I–VII ionic semiconductor. γ-CuCl with zinc blende structure, which is stable below 407 °C, has a direct bandgap with E_g = 3.395 eV at 4 K [1]. It has a high exciton binding energy of ~190 meV [2], which is higher than both GaN (25 meV) [3] or ZnO (60 meV) [4]. CuCl has been investigated for some time for its application to optoelectronic devices [5–8]. The high binding energy gives the possibility of stable, room temperature, UV emission which, together with high biexciton binding energies, enables optoelectronic effects such as bistability and four-wave mixing with the potential for new short wavelength devices [9,10].

In order to use CuCl in optoelectronic devices it has to be deposited in thin films or in arrays of nanoparticles. Thin films of CuCl have been deposited by thermal evaporation, molecular beam epitaxy, and magnetron sputtering [11–13]. Arrays of nanoparticles of Cu halides have been produced in a matrix of glass, silicon, or organic compounds by gas or liquid phase methods, typically in a three dimensional form [6,14–17] but the size of the crystallites has been difficult to control. Atomic layer deposition (ALD) has recently been shown to be able to produce two-dimensional nanocrystalline arrays of CuCl on substrates and the size and distribution of the crystallites could be controlled by the parameters of the deposition process [18,19]. These films have shown the crystal structure of γ-CuCl and demonstrated the luminescent and optical characteristics typical of CuCl. One important feature of CuCl is that it is sensitive to moisture in its environment. Films of CuCl will hydrolyse to oxy- or hydroxy-halides after a short time and they need hermetic protection for long-term stability. Successful encapsulation by spun-on organic materials such as polysilsesquioxanes and cycloolefin copolymers has proved successful while plasma-enhanced chemical vapour deposition of SiO_2 has not provided adequate encapsulation [20]. These spun-on techniques required short-term exposure of the CuCl to the atmosphere so there is still the possibility of some degradation before encapsulation. In addition, the relative thickness and probable lack of uniformity of spun-on films will be a drawback in device construction.

ALD is a chemical vapour deposition technique characterised by its ability to controllably deposit ultrathin layers with extreme uniformity. In ALD, the substrate is exposed sequentially to pulses of reactant gases or vapours and each pulse forms an additional chemisorbed molecular layer on it. Between the reactant pulses, an inert gas is used as a purge gas for removing all the excess precursor molecules that have not chemisorbed or undergone exchange reactions with the surface groups, and removing the reaction byproducts [21]. A single sequence of precursor and purge pulses is known as an ALD cycle. Initially, the film formation may proceed by the nucleation and growth of individual nanocrystallites, whose size and density varies according to the number of ALD cycles used in the process [22]. The details of this nucleation process are dependent on the chemical interaction between the precursors and the initial surface chemical state of the substrate, which is affected by its pretreatment. This process of controlled nucleation provides a possible method of producing plasmonic structures in copper halides in a much more repeatable way. In the previous ALD work [18,19], the Cl precursor used, i.e., a solution of HCl in butanol, had limitations with respect to questionable stability of the vapour pressure and possible bi-reactions with butanol itself.

In this publication, we report on the sequentially pulsed chemical vapour deposition of CuCl using an alternative Cl precursor which circumvents the problems of the one used in previous ALD of CuCl. We study the change in the distribution of nanocrystallites as deposition progresses. We characterise the structural, optoelectronic and chemical properties of the film and explore the use of ultrathin diffusion barrier layers deposited in situ to provide environmental protection.

2. Materials and Methods

The films were deposited simultaneously on quartz glass, Si, and soda-lime glass substrates by thermal pulsed chemical vapour deposition using a TFS 200 ALD system (Beneq Oy, Espoo, Finland). The deposition temperature and chamber pressure were 125 °C and 2 mbar, respectively, with 100, 200, 500, or 1000 deposition cycles. The precursor for copper was [Bis(trimethylsilyl)acetylene]-(hexafluoroacetylacetonato)copper(I) (Sigma-Aldrich, St. Louis, MO, USA), abbreviated to CuBTMSA for convenience heated to 80 °C, and the precursor for chlorine was pyridine hydrochloride (Sigma-Aldrich, ≥98%) heated to 50 °C. One CuCl deposition cycle was defined by the following sequence. Cu pulse (2 s) → N_2 purge (4 s) → Cl pulse (3 s) → N_2 purge (6 s). Nitrogen was also used as a carrier gas. Some samples were protected against oxidation and atmospheric moisture by an Al_2O_3 capping layer. This capping layer was deposited in situ by ALD from trimethylaluminium (TMA, Strem Chemicals, Newburyport, MA, USA; electronic grade) and ozone (ozone generator, BMT Messtechnik, Stahnsdorf, Germany; 8 g/h output) at 150 °C. One Al_2O_3 ALD deposition cycle was defined by the following sequence. TMA pulse (0.1 s) → N_2 purge (3s) → O_3 pulse (0.3 s) → N_2 purge (3 s). Ozone was used as the oxidiser here because of the fear that using water might hydrolyse the CuCl. The capping layer consisted of 47 ALD cycles giving an Al_2O_3 thickness of approximately 5 nm. The deposition chamber was connected to a nitrogen glove box with moisture and oxygen content values < 0.1 ppm. The samples were stored in the glove box and only removed for the time necessary for analyses except for those capped samples deliberately exposed to the atmosphere.

ALD is characterised by growth saturation with increasing precursor dose and the presence of an ALD temperature window. In the process described here, the full range of deposition parameters has not yet been explored with these precursors to definitively prove that the CuCl deposition was purely an ALD process. Therefore, the process can only strictly be called sequentially pulsed chemical vapour deposition. We consider, however, that in comparing the growth of different samples it is in order to refer to the number of "ALD cycles" as a shorthand descriptor rather than saying "sequentially pulsed precursor-purge cycles".

The CuCl was deposited on various types of substrate: (i) silicon wafers (100) (P/Boron doped, thickness = 525 ± 25 μm, Ω = 10–30 Ω·cm); (ii) soda-lime glass; and (iii) quartz glass Herasil 102 (Heraeus, Hanau, Germany). Before the deposition the samples of size approximately 1×1 cm^2 were cleaned in an ultrasonic bath: 5 min in acetone, 5 min in isopropanol, and dried with nitrogen. Hydrofluoric acid was used to etch native oxide from the Si wafer surfaces. Transparent samples (soda-lime glass and quartz glass) were coated with Kapton tape on the back side to allow deposition only on one side. The Kapton tape was removed prior to optical measurements.

X-ray photoelectron spectroscopy (XPS) was carried out using an ESCALAB 250Xi apparatus (ThermoFisher Scientific, Waltham, MA, USA) with monochromated Al Kα radiation (1486.6 eV) to determine the chemical composition of CuCl thin films. All samples were measured at two spots at a takeoff angle of 90° in 10^{-8} mbar vacuum at 20 °C. An electron flood gun was used to compensate for charges on sample surfaces. The spectra were referenced to C–C at 284.8 eV. The spot size was 650×650 μm^2 and the pass energy was 50 eV for the survey and 20 eV for the high resolution scan. The elemental composition was estimated from the survey spectra using Avantage ver. 5.938 software. The measurements on sample Q8 after atmospheric exposure were carried out using an Axis Supra instrument (Kratos Analytical, Manchester, UK) under slightly different conditions; spot size 700×300 μm^2 and the pass energy was 120 eV for the survey and 20 eV for the high resolution scan. The analysis was done by CasaXPS software ver. 2.3.19PR1.0. The high-resolution spectra were fitted using XPSPEAK software (ver. 4.1) with Shirley and/or linear background.

Grazing incidence X-ray diffractometry (GIXRD) was used to measure the crystalline properties of the CuCl. The measurements were taken with a SmartLab diffractometer (Rigaku, Tokyo, Japan) at grazing incidence angle $\alpha = 0.2°$ using a copper X-ray tube (1.542 Å).

Normal-incidence reflectance was measured in the Vis–UV range using an Avantes 2048 pixel spectrometer (Avantes BV, Apeldoorn, The Netherlands; spectral range from 1.9 to 5.1 eV), equipped with a fibre reflectance probe [23]. The angles of incidence covered the range of 0.2°–1.7° and the nearly circular spot had a measured diameter of 0.8 mm for the 400 μm detection fibre. The small measurement spot allowed us to test the in-plane homogeneity of the samples. Bare substrates were used to collect reference signals.

The photoluminescence (PL) data were obtained with using a LabRam HR Evolution (HORIBA Scientific, Kyoto, Japan) 0.8 m focal length single-stage spectrometer with CW HeCd laser excitation at 325 nm (3.8 eV). The mirror lens collected signals from approximately a 1 μm circular spot; the excitation power was kept low enough to prevent heating of the layers (~200 W cm^{-2}). The PL spectra were recorded with a Peltier-cooled back-illuminated UV-sensitive CCD detector Synapse 2048×512. The instrument spectral range sensitivity has been corrected by the Intensity Correction System (HORIBA Scientific, Kyoto, Japan) delivered with the system. All reflectance and PL measurements were performed at room temperature (300 K).

3. Results and Discussion

3.1. Deposition

Table 1 shows the deposition conditions for quartz glass samples.

The growth habit of the films on all substrates showed individual nuclei growing in number and size as the number of cycles increased. Figure 1 shows SEM images of the uncapped films on glass for 100, 200, 500, and 1000 ALD cycles (abbreviated to c. hereafter) deposited on soda-lime glass. The films on Si substrates were similar.

Table 1. Deposition parameters for films on quartz substrates.

Sample No.	No. of ALD Cycles	Capping Layer
Q1	100	no
Q5	200	no
Q7	500	no
Q10	1000	no
Q2	100	yes
Q6	200	yes
Q8	500	yes

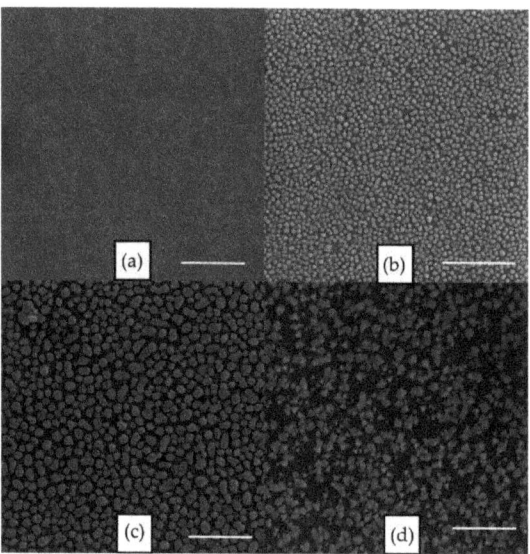

Figure 1. SEM images of CuCl deposited on soda-lime glass: (**a**) 100 c.; (**b**) 200 c.; (**c**) 500 c.; and (**d**) 1000 c. Scale bars represent 1 µm.

As the number of cycles increased, the diameter of the crystallites also increased. There was an initial nucleation phase and then the surface was covered in an array of crystallites. The areal density of the crystallites showed a decrease with the number of cycles; some form of agglomeration or ripening was evident. Dimensional change of the crystallites can take place by two processes: (i) Coalescence, where there is poor adhesion between the nanocrystallites and the substrate and they are mobile enough to move across it and to coalesce and (ii) Ostwald ripening, in which there is strong adhesion of the nanocrystallites to the surface and material transfer takes place by the diffusion of atoms across the surface. In this case, there is no significant coalescence of nanocrystallites therefore there is an Ostwald ripening process taking place. Initially, the crystallites had somewhat "liquid-like" shapes whereas the crystallites in the 1000 c. films had a more facetted appearance. This change in habit was accompanied by differences in the chemical composition of the films as discussed later. It is possible that this was caused by the longer deposition time required for the thicker film giving more time for reconfiguration of the crystallites. The size of the crystallites in the CuCl films increases with the number of cycles as shown in Figure 2a, while Figure 2b shows this behaviour in comparison with that reported previously using different precursors [19].

Figure 2 demonstrates that the rate of increase of crystallite diameter is similar with both the current and the previous precursors, once the initial nucleation has taken place, and that the growth on crystalline and noncrystalline substrates is similar. This similarity is not unexpected since the reactant is HCl in both cases; here it is derived by decomposition of pyridine hydrochloride, previously the HCl vapour came from an HCl/butanol solution. The nucleation behaviour is somewhat different: with PyrHCl there is slower nucleation. However, with the HCl/butanol precursor the nucleation behaviour depended on the growth cycle parameters so further exploration of the deposition parameters with PyrHCl may also show differences in nucleation behaviour.

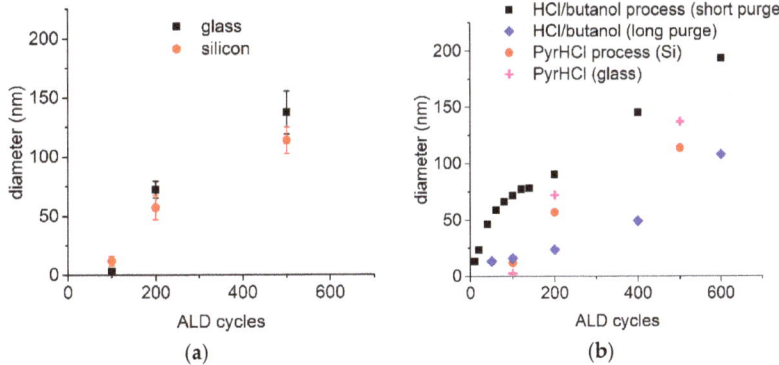

Figure 2. (a) Crystallite diameter vs. number of atomic layer deposition (ALD) cycles using pyridine hydrochloride (PyrHCl) precursor and (b) comparison with previous results using HCl/butanol as precursor (HCl). ● PyrHCl–Si, + PyrHCl–glass, ■ HCl (short purge), ◆ HCl (long purge). In (b) error bars have been omitted for clarity.

3.2. Crystal Structure

Films for XRD analysis were deposited on quartz glass both without a capping layer and with a capping layer of ~5 nm of aluminium oxide deposited in situ in order to prevent the films undergoing hydrolysis in atmospheric moisture. The crystal structure was determined by X-ray diffraction (XRD) using GIXRD. The results are shown in Figure 3a,b. The films show the zinc blende crystal structure of γ-CuCl with no evidence of any other crystalline phases. The films with an Al_2O_3 capping layer also showed the same structure with small additional broad peaks at ~36° and 45° (marked with asterisks). These do not fit $CuCl_2$ (PDF 9001506), CuO (PDF 1011148), Cu_2O (PDF 1010926), or other likely compounds. It is not clear what material they represent. The crystallites showed largely random orientation although with some preferential (220) orientation for thinner films. For randomly oriented films the intensity ratio $I_{220}/I_{111} \approx 0.60$ and the ratio here is much higher. However, it must be borne in mind that because of the glancing angle X-ray incidence, any preferential orientation is not parallel to the substrate. From these results we can conclude that the bulk of the film is γ-CuCl.

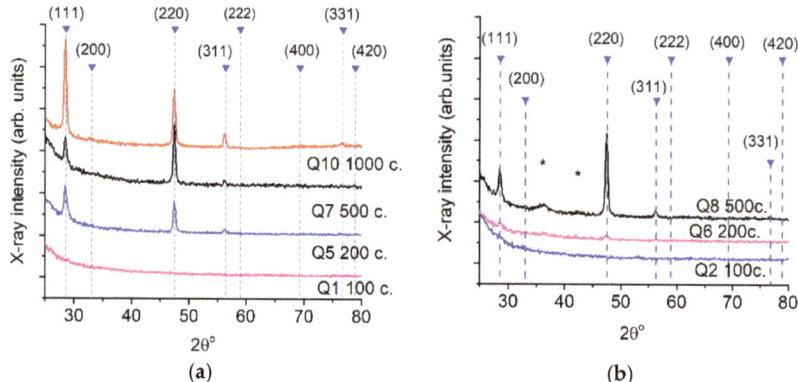

Figure 3. XRD spectra of (**a**) uncapped and (**b**) capped CuCl films on quartz substrates. Unidentified peaks marked with asterisks. Note: The "missing" CuCl peaks ((200), (222), (400), and (420)) have low intensities compared to the (111) peak for randomly oriented material (PDF 1010991).

3.3. Chemical Composition

The chemical composition was investigated by XPS. The scans were taken before and after Ar sputter-cleaning of the substrate to investigate the extent of surface contamination. Table 2 shows the analysis of survey scans of the deposited material for the unsputtered samples while Table 3 shows the relative concentrations of Cu and Cl after Ar sputter cleaning for 150 s. The XPS analysis was not performed in situ so there will be significant amounts of adventitious carbon from environmental contamination between deposition and analysis for the unsputtered films. The sputtered samples show significantly reduced C 1s signal. There is evidence, however, of the presence of carbon as a residue from the organic precursors (see below). There is also some fluorine content in all films which must come from the CuBTMSA precursor. There is significant uncertainty in the Cu/Cl ration due to the dispersed crystallites which allow a large signal from the substrate material to be detected and because the relative sensitivity factor for the XPS quantification for the Cl 2p peak is much lower than for the Cu 2p peaks by a factor of ~11. Nevertheless, for the unsputtered samples the Cu/Cl ratio is lower than the expected 1/1 ratio for CuCl; however, the sputter-cleaned samples show close to the stoichiometric ratio. The low CuCl ratio for the unsputtered samples arises from the different composition of the surface layer.

Table 2. Element concentration obtained from X-ray photoelectron spectroscopy (XPS) survey scan (not Ar-sputtered).

Sample		Cycles	Concentration of Elements (at.%)							
			Cu 2p	Cl 2p	C 1s	Si 2p	Al 2p	O 1s	F 1s	Cu/Cl
No cap	Q1	100	0.5	3.1	19.6	29.9	0.0	46.6	0.4	0.1
	Q5	200	1.9	2.7	12.9	31.3	0.0	50.5	0.7	0.7
	Q7	500	1.4	2.7	15.7	31.1	0.0	48.6	0.5	0.5
	Q10	1000	3.8	10.2	50.7	12.8	0.0	21.0	0.6	0.4
Capped	Q2	100	0.3	0.8	22.1	8.4	21.9	46.3	0.3	0.3
	Q6	200	0.2	0.5	23.9	6.0	23.2	45.7	0.4	0.4
	Q8	500	0.2	0.7	25.9	5.5	24.0	43.5	0.3	0.3

Table 3. Relative concentrations of Cu and Cl obtained from XPS (after Ar sputtering).

Sample	Cycles	Relative Concentration of Cu and Cl		
		Cu 2p	Cl 2p	Cu/Cl
Q7	500	55	45	1.2
Q10	1000	49	51	1.0
Q8 (capped)	500	47	53	0.9

High-resolution scans of the Cl 2p peak from the uncapped samples with peak fittings are shown in Figure 4 for the unsputtered and sputter-cleaned samples. For Q1, Q5, and Q7 the Cl 2p peak is well fitted by one 2p doublet consisting of $2p_{3/2}$ and $2p_{1/2}$ components separated by 1.6 eV. Figure 4a is representative of these. The binding energy values are consistent with those given for copper chlorides although they cannot easily distinguish between Cu(I) and Cu(II) chloride. In general, the $CuCl_2$ should appear at a higher binding energy [24–26]. Taking into account the XRD results which show only CuCl we conclude that peak at ~199.0 eV is CuCl. There is no significant ClO_x formation: this would give rise to $2p_{3/2}$ peaks at energies approximately 206–208 eV [27], and there is no evidence of these in the Cl 2p spectra. The peak details are given in Table 4. For sample Q10, if a single peak fitting is used, the fitting error is significantly higher. However, the measured Q10 signal can be well fitted with two 2p components separated by 0.6 eV (Figure 4b). This is consistent with the emergence of two different bonding environments for Cl. The lower energy peak is characteristic of $HCl \cdot nH_2O$ (198.4 eV) [28] and is also consistent with CuCl bound to an organic ligand [29]. The peak area ratio between this component and the one at slightly lower energy is 1.4 indicating significant surface contamination. Figure 4c shows the Cl 2p peak for Q7 after Ar sputter-cleaning. For the Ar-sputtered Q10 sample, again the best fit is with two doublets. In this case, the additional peak appears at higher energy, consistent with the existence of an organic chloride [28,30].

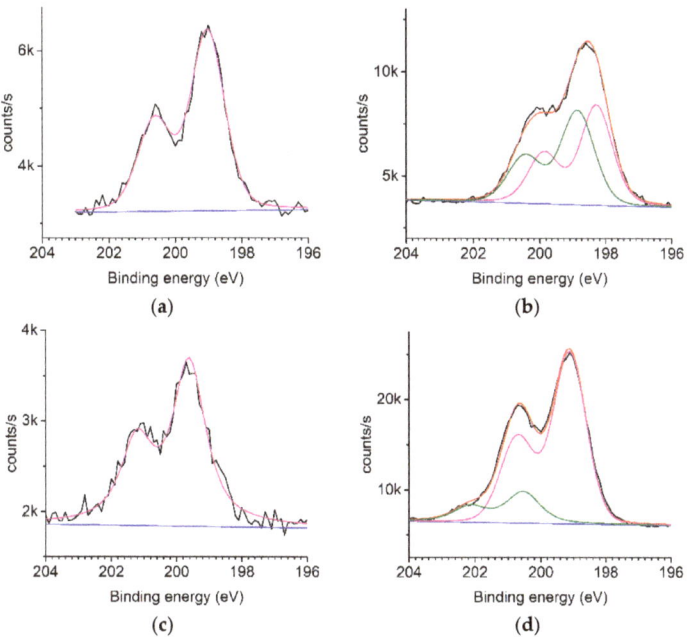

Figure 4. XPS Cl 2p spectra for uncapped quartz samples showing fitted peaks: (**a**) Unsputtered Q7 500 c.; (**b**) unsputtered Q10 1000 c.; (**c**) Ar-sputtered Q7 500 c.; and (**d**) Ar-sputtered Q10 1000 c.

Table 4. Components of Cl 2p XPS peaks.

Sample	Cycles	Peak	2$p_{3/2}$ Binding Energy (eV)	Area P2/P1	Peak FWHM (eV)
Q7, unsputtered	500	P1	199.0	–	1.2
Q7, Ar-sputtered	500	P1	199.6	–	1.2
Q10, unsputtered	1000	P1	198.9	1.0	1.2
		P2	198.3		1.1
Q10, Ar-sputtered	1000	P1	199.1	0.2	1.2
		P2	200.5		1.4
Q8 (capped), Ar-sputtered	500	P1	199.3	–	2.2

Figure 5 shows the high resolution $2p_{3/2}$ and $2p_{1/2}$ raw data scans scan of Cu and the decomposition of the $2p_{3/2}$ peaks into individual components. Q7 is representative of the other samples with less than 1000 c. deposition. Table 5 shows the peak positions and full width half maxima (FWHM).

It is difficult to separate the Cu^0, Cu^+, and Cu^{2+} peaks solely by their binding energies but taking into account the XRD spectra, peak P1 is assigned to CuCl which has been previously seen at 932.4–932.6 eV [31,32]. Peak P2 (~934 eV) could be indicative of CuO, $CuCl_2$, or Cu with an organic or organo-chlorate ligand [33–36]. However, the presence of Cu^{2+} species should be indicated by an obvious shake-up satellite peak at ~940 eV [37] and there is no evidence of this in Q7 both with and without sputtering (Figure 5a,b). Neither is there evidence for chlorate bonding in the Cl 2p spectra. Therefore, Q7 contains only Cu^+ bonds and P2 is ascribed to Cu(I) organic bonding. The amount of the Cu organic contamination in both Q7 and Q10 is clearly less after Ar sputtering (Figure 5d,f). Figure 5a,e, Q10 (1000 c., unsputtered), shows a large satellite peak at the expected energy for Cu^{2+} at ~940 eV together with a peak at 936.7 eV, P3, which may be evidence for some fluoride bound to organic material [38]. However, the Q10 Ar-sputtered sample, Figure 5f, shows the same three peaks as for the Q10 unsputtered sample with no evidence of a satellite peak. Therefore, it is not clear which of the peaks is related to the Cu^{2+} satellite peak. It may be that the relatively lower intensity of P3 after Ar sputtering gives rise to a much smaller satellite peak which is not visible in Figure 5f.

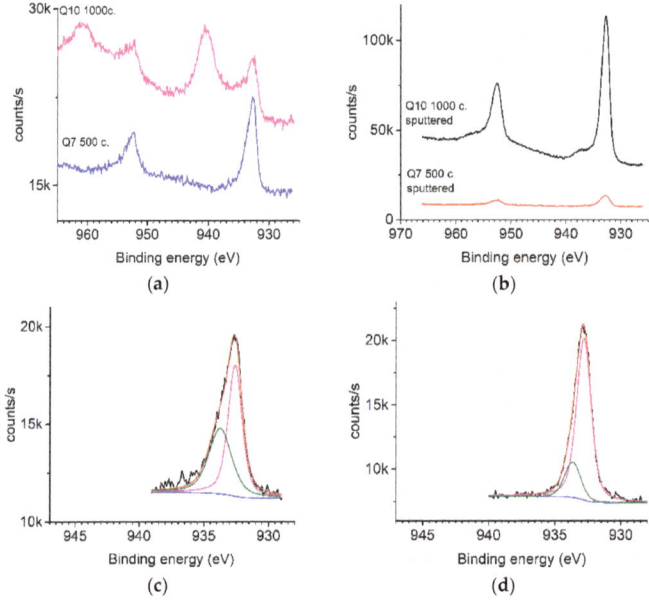

Figure 5. *Cont.*

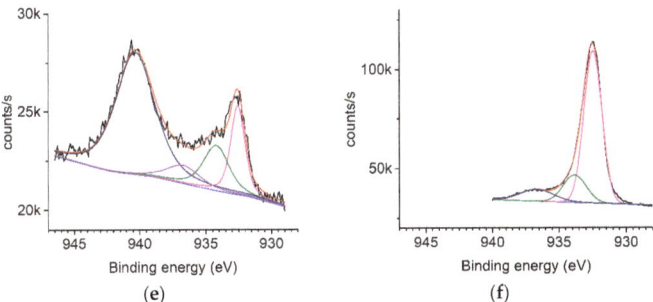

Figure 5. XPS Cu 2p peaks, uncapped quartz samples: (**a**) Q7 500 c. and Q10 1000 c., unsputtered; (**b**) Q7 500 c. and Q10 1000 c., Ar-sputtered; (**c**) decomposed $2p_{3/2}$ peak, Q7, unsputtered; (**d**) decomposed $2p_{3/2}$ peak, Q7, Ar-sputtered; (**e**) decomposed $2p_{3/2}$ peak, Q10, unsputtered; and (**f**) decomposed $2p_{3/2}$ peak, Q10, Ar-sputtered.

Table 5. Peak parameters for Cu 2p XPS.

Sample	Cycles	Peak Cu $2p_{3/2}$	Binding Energy (eV)	FWHM	Assignment
Q7, unsputtered	500	P1	932.6	1.3	CuCl
		P2	933.7	2.1	Cu(I) org.
Q7, Ar-sputtered	500	P1	932.8	1.3	CuCl
		P2	933.7	2.1	Cu(I) org.
Q10, unsputtered	1000	P1	932.6	1.3	CuCl
		P2	934.2	2.3	Cu(I) org.
		P3	936.7	3.0	organofluoride
		S	940.3	3.5	Cu^{2+}
Q10, Ar-sputtered	1000	P1	932.5	1.5	CuCl
		P2	933.8	2.0	Cu(I) org.
		P3	936.7	3	organofluoride
Q8, Ar-sputtered	500	P1	933.2	1.8	CuCl
		P2	934.7	2.1	Cu(I) org.

Observation of the O 1s spectra yields further information. The spectra for Q7 and Q10 before and for Q10 after Ar sputtering are shown in Figure 6. The samples are dominated by the peak at 533.3 eV due to SiO_2 [39] from the quartz since the substrate coverage is not complete. Both unsputtered samples, Figure 6a,b, also show a peak at ~534 eV which is probably O in an organic ligand indicating some residual CuBTMSA precursor material in the film. In addition, Figure 6b shows a large additional peak at 535.8 eV, which is at an energy which has been ascribed to an O atom linked to a CF containing ligand [30]. This again is likely due to some partially decomposed CuBTMSA precursor. Figure 6c is representative of all the sputtered samples and shows only the SiO_2 peak indicating that the oxygen bonded to organic material is only on the surface.

When the Cl 2p, Cu 2p, and O 1s results are taken together with the XRD results, it is clear that the bulk of the films consists of γ-CuCl with a certain amount of organic contamination from the precursor molecules. This is confirmed by the presence of significant carbon content even in the Ar-sputtered samples. The unsputtered samples show, as is to be expected, greater organic contamination, including from F-containing organic fragments and HCl from the CuBTMSA and PyrHCl precursors, respectively. The exact nature of the bonding is the subject of further investigation.

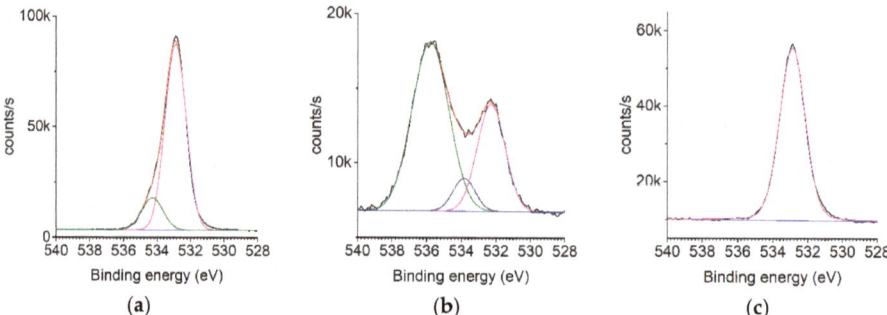

Figure 6. Decomposition of O 1s peaks for uncapped quartz samples: (**a**) Q7 500 c., unsputtered; (**b**) Q10 1000 c., unsputtered; and (**c**) Q10 1000 c., Ar-sputtered.

Thin uncapped CuCl films will hydrolyse after approximately one day. Figure 7 and Tables 3–5 show measurements on the sputter cleaned sample Q8 (500 c., Al$_2$O$_3$ capped) after exposure to normal atmosphere for approximately 3 weeks. The curves and tabulated data are very similar to those of the uncapped samples with the same number of growth cycles kept in an inert atmosphere (Figures 4c and 5e). They show no evidence of degradation, indicating that 5 nm of oxide capping layer provides an effective barrier to hydrolysis of the films.

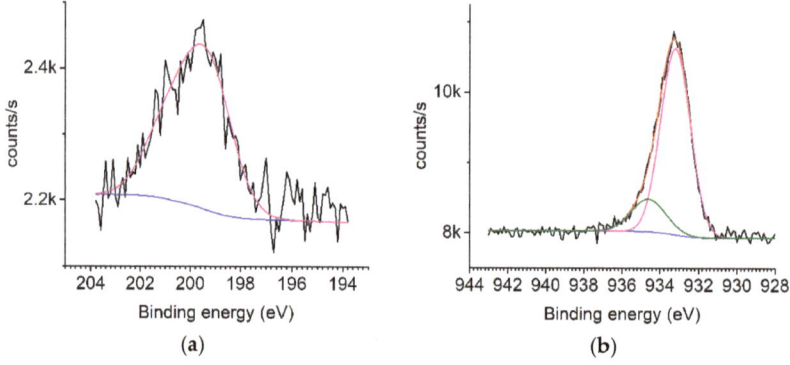

Figure 7. Decomposition of (**a**) Cl 2p and (**b**) Cu 2p$_{3/2}$ peaks of Q8 (500 c. capped, Ar-sputtered) after approximately 3 weeks in normal air.

3.4. Photoluminescence

The photoluminescence signals obtained from sample Q10 (1000 c., uncapped) were easily measurable at room temperature. Shown in Figure 8 is the experimental spectrum, decomposed into three Gaussian bands located on a linear background. The sum fits the measured data within background noise. The band parameters are listed in Table 6. We use band area as the measure of its strength. The location of the PL maximum is very close to the position of the strongest PL2 band, coinciding with the position of the absorption peak PL2 (3.248 eV) of Table 6 within experimental uncertainty. The weaker but discernible high-energy PL3 band at 3.308 eV lies 0.06 eV above PL2. The lower energy band PL1 is situated at 3.225 eV, approximately 0.10 eV lower than PL2. These results are similar to those observed on CuCl films obtained by thermal evaporation [40] and magnetron sputtering [41]. Those showed that at low temperature the emission peak could be resolved into three components which broadened into one composite peak at room temperature. The main peak was identified as the Z_3 free exciton peak observed at energy 3.227 eV at 15 K. This peak shifted to higher energy of 3.243 eV at room temperature, a shift of approximately 0.016 eV [42]. This peak is almost at the same energy as the main peak reported here (3.248 eV). The peak shown here can also be decomposed into three components. By comparison with these results and allowing for the shift with temperature, the main peak indicating the most probable radiative channel is ascribed to the Z_3 exciton. The lower energy peak we see at 3.225 eV (385.4 nm) can be ascribed to the I_1 peak due to an exciton bound to an impurity, probably a Cu vacancy [43]. The higher energy peak at 3.308 eV (375.8 nm) can be ascribed to the $Z_{1,2}$ exciton. PL mapping across the sample shown in Figure 9 (32 × 32 µm², step 4 µm) shows intensity varying by a factor of approximately 2, due to local variations of the effective film thickness. No changes in peak position, width or shape were observed.

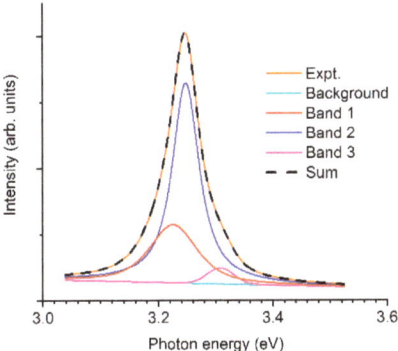

Figure 8. Photoluminescence emission intensity vs. photon energy showing resolved peaks for Q10 (1000 c. uncapped sample). The dashed line is the fitted peak.

Table 6. Best-fit parameters of the Gaussian bands used in modelling the PL spectrum of Figure 8.

Band	Strength (a.u.)	Position (eV)	Width (eV)
PL1	5.8 ± 0.3	3.225 ± 0.010	0.097 ± 0.010
PL2	13.1 ± 0.3	3.248 ± 0.010	0.056 ± 0.010
PL3	0.7 ± 0.3	3.308 ± 0.010	0.054 ± 0.02

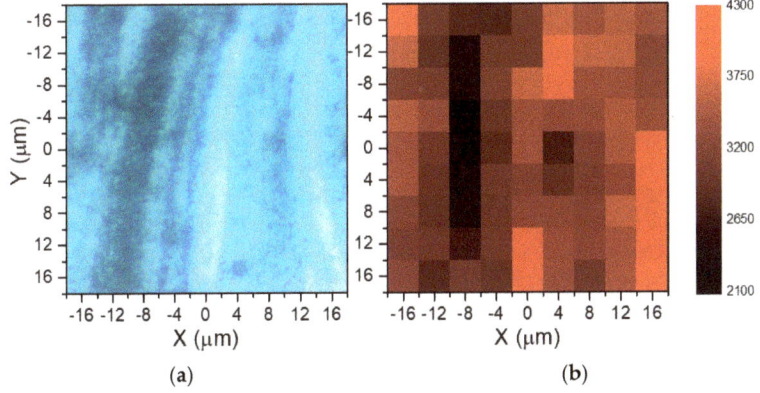

Figure 9. Uncapped quartz sample Q10, 1000 c. (a) White light photography and (b) integral PL intensity (arb. units) over the same area of the substrate with 4 × 4 µm² resolution.

3.5. Reflectance

The measured reflectance spectrum of the thickest uncapped CuCl film (Q10 1000 c.) is shown in Figure 10a. It shows a significant increase of the reflected intensity compared to the bare substrate, which is indicative of a larger optical density of the film. In addition, a fairly strong spectral structure with maximum intensity at about 3.3 eV has apparently at least two narrow components. The latter are better seen in the twice differentiated spectra plotted in Figure 10b. The differentiation enhances the sharper structures and suppresses the flat background. Based on the narrow structures in the derivative spectra, we have fitted the measured reflectance with a model consisting of a flat background dielectric function, with two superposed Gaussian absorption bands. The model lineshape of Figure 10a,b represents the best-fit results with the (fitted) value of the film thickness of 29 nm, and the background value of the real part of its dielectric function of 3.2. The resulting parameters of the bands are listed in Table 7. The strength parameter is proportional to the area below the absorptive (imaginary) part of the dielectric function, having the units of eV; however, we are interested in relative values and use arbitrary units here.

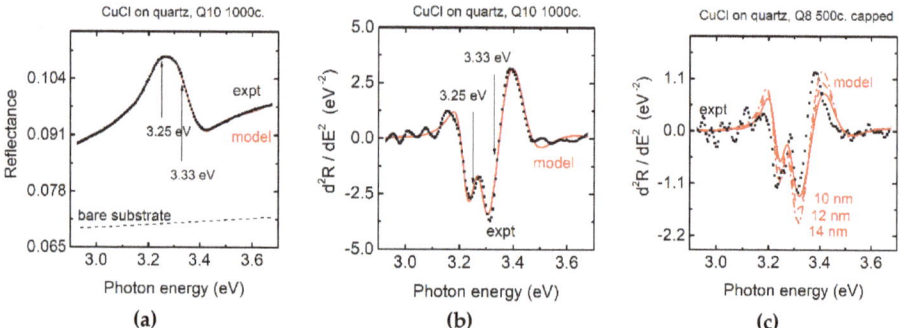

Figure 10. (a) Normal incidence reflectance spectra of the uncapped, 1000 c. sample; the positions of two prominent Gaussian bands are indicated by arrows, bare substrate (dashed line); (b) second derivative; and (c) second derivative of the reflectance spectrum from the capped, 500 c. sample. Measured data (symbols), model lineshapes (red solid and dashed lines).

Table 7. Best-fit parameters of the Gaussian bands used in modelling the reflectance spectra of Figure 10.

Band	Strength (a.u.)	Position (eV)	Width (eV)
R1	0.03 ± 0.01	3.25 ± 0.01	0.06 ± 0.02
R2	0.23 ± 0.01	3.33 ± 0.01	0.14 ± 0.02

The two Gaussian bands are 0.08 eV apart, which is in a good agreement with the distance of 0.06 eV seen in the absorption spectra of Z_3 (3.23 eV) and $Z_{1,2}$ (3.308 eV) excitons at in the PL results and those seen at 4 K [44]. In addition, the significantly larger strength of the upper (3.30 eV) band is clearly seen in the reflectance results, in contrast to the PL intensities. This is similar to the relative intensities seen in direct optical absorption measurements [18,45]. We have also observed a weak and broad luminescence band centred at about 2.7 eV, with the maximum signal below 4% of the main PL band; its high energy tail forms the smooth background seen in Figure 7. This may be due to radiative recombination of defect states in the CuCl crystallites, as well as other species present in the deposited layer.

A film of the sample Q8 (500 c. with an Al_2O_3 capping layer) displays a reduced intensity excitonic structure in the differentiated reflectance, see Figure 10c. The results are compatible with reduced effective film thickness of approximately 12 nm. In addition, the upper lineshape seems to be narrower than that of the thicker sample Q10, probably due to the smaller dispersion of the crystallite sizes. It can be seen that reflectance measurements are a useful way to identify the presence of CuCl in thin films. This method would be particularly useful where the film is deposited on a non-UV transparent substrate.

The reflectance of sample Q8 was remeasured after approximately three weeks in normal atmosphere. A comparison between the two measurements is shown in Figure 11. There is no significant change to the excitonic absorption confirming the XPS results that 5 nm of ALD Al_2O_3 provides an effective protection layer for the CuCl. This is consistent with published work showing that ALD Al_2O_3 constitutes a good moisture diffusion barrier when applied to other materials [46].

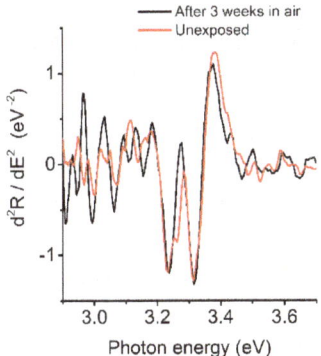

Figure 11. Comparison of reflectance curves (2nd derivative) for aluminium oxide-capped sample Q8 before and after approximately 3-week exposure to ambient air.

4. Conclusions

It has been demonstrated that CuCl can be deposited by sequentially pulsed vapour deposition process using the precursors [Bis(trimethylsilyl)acetylene](hexafluoroacetylacetonato)copper(I) and pyridine hydrochloride and that an in situ ALD capping layer of aluminium oxide is an effective barrier for preventing atmospheric degradation. The crystal structure has been shown to be the zinc blende-structure γ-CuCl by XRD. The crystallites become more facetted as the film thickness increases. The bulk of films shows only some organic contamination from incomplete precursor reaction, however, there is heavier organic contamination on the surface, again from unreacted precursor molecules. The characteristic photoluminescence behaviour of CuCl has been shown with emissions from the $Z_{1,2}$, Z_3 and bound excitons. The chemical composition was investigated by XPS: the 200 c. and 500 c. films show surface layers with predominantly Cu^+ bonding from CuCl with some organic contamination. The thicker films with 1000 deposition cycles have significant Cu^{2+} content from CuF_2 or F containing organic fragments. Optical reflectance measurements have shown that the characteristic exciton absorptions can be detected, at similar energies to optical absorption features measured by transmission. This enables exciton absorption measurements to be carried out on nontransparent samples.

The overall results show that this is a method for deposition of nanocrystalline arrays suitable for further investigation for the development of new optoelectronic structures and devices. Further studies are also required to clarify the difference in chemical composition between the surface layers and the bulk of the film.

Author Contributions: Conceptualization, D.C.C; Methodology, D.C.C.; Formal Analysis, T.H., R.K., and D.C.C.; Investigation, R.K., T.H., O.C., K.K., R.Z., J.P., and J.M.M.; Resources, R.Z., J.P., and J.M.M.; Writing–Original Draft Preparation, T.H. and J.H.; Writing–Review & Editing, D.C.C., J.H., R.Z., J.M.M., R.K., T.H., K.K., and O.C.; Visualization, D.C.C. and J.H.; Supervision, D.C.C., J.H., and J.M.M.; Project Administration, D.C.C. and R.K.; Funding Acquisition, D.C.C. and J.M.M.

Funding: This work was financially supported by the Czech Science Foundation (No. 17-02328S), the Ministry of Youth, Education and Sports of the Czech Republic (Nos. LM2015082, LQ1601, LO1411 (NPU I), and CZ.02.1.01/0.0/0.0/16_013/0001829), and the European Regional Development Fund (project CZ.1.05/2.1.00/03.0086).

Acknowledgments: The authors acknowledge the support of COST action MP1402 HERALD in the course of this work.

Conflicts of Interest: The authors declare no conflicts of interest.

References

1. Soga, M.; Imaizumi, R.; Kondo, Y.; Okabe, T. A method of growing CuCl single crystals with flux. *J. Electrochem. Soc.* **1967**, *114*, 388–390. [CrossRef]
2. Nakayama, M.; Ichida, H.; Nishimura, H. Bound-biexciton photoluminescence in CuCl thin films grown by vacuum deposition. *J. Phys. Cond. Matter* **1999**, *11*, 7653–7662. [CrossRef]
3. Monemar, B. Fundamental energy gap of GaN from photoluminescence excitation spectra. *Phys. Rev. B* **1974**, *10*, 676. [CrossRef]
4. Ryu, Y.R.; Lee, T.S.; White, H.W. Properties of arsenic-doped-type ZnO grown by hybrid beam deposition. *Appl. Phys. Lett.* **2003**, *83*, 87–89. [CrossRef]
5. Alam, M.M.; Lucas, F.O.; Danieluk, D.; Bradley, A.L.; Daniels, S.; McNally, P.J. Temperature dependent photoluminescence of nanocrystalline γ-CuCl hybrid films. *Thin Solid Films* **2014**, *564*, 104–109. [CrossRef]
6. Alam, M.M.; Lucas, F.O.; Danieluk, D.; Bradley, A.L.; Rajani, K.V.; Daniels, S.; McNally, P.J. Hybrid organic–inorganic spin-on-glass CuCl films for optoelectronic applications. *J. Phys. D Appl. Phys.* **2009**, *42*, 225307. [CrossRef]
7. Göbel, A.; Ruf, T.; Cardona, M.; Lin, C.T.; Wrzesinski, J.; Steube, M.; Reimann, K.; Merle, J.C.; Joucla, M. Effects of the isotopic composition on the fundamental gap of CuCl. *Phys. Rev. B* **1998**, *57*, 15183. [CrossRef]
8. Mitra, A.; O'Reilly, L.; Lucas, O.F.; Natarajan, G.; Danieluk, D.; Bradley, A.L.; McNally, P.J.; Daniels, S.; Cameron, D.C.; Reader, A.; Martinez-Rosas, M. Optical properties of CuCl films on silicon substrates. *Phys. Status Solidi B* **2008**, *245*, 2808–2814. [CrossRef]
9. Grun, J.B.; Hönerlage, B.; Levy, R. Dynamics of optical nonlinearities and bistability in 3-level systems (CuCl). *Phys. Scr.* **1986**, *T13*, 184–188. [CrossRef]
10. Gomes, M.; Kippelen, B.; Hönerlage, B. Time, intensity and energy dependence of four-wave mixing processes in CuCl. *Phys. Stat. Solidi B* **1990**, *159*, 101–106. [CrossRef]
11. Natarajan, G.; Daniels, S.; Cameron, D.C.; O'Reilly, L.; Mitra, A.; McNally, P.J.; Lucas, O.; Rajendra Kumar, R.T.; Reid, I.; Bradley, A.L. Stoichiometry control of sputtered CuCl thin films: Influence on ultraviolet emission properties. *J. Appl. Phys.* **2006**, *100*, 033520. [CrossRef]
12. Kawamori, A.; Edamatsu, K.; Itoh, T. Growth of CuCl nanostructures on CaF_2(111) substrates by MBE—Their morphology and optical spectra. *J. Cryst. Growth* **2002**, *237–239*, 1615–1619. [CrossRef]
13. O'Reilly, L.; Lucas, O.F.; McNally, P.J.; Reader, A.; Natarajan, G.; Daniels, S.; Cameron, D.C.; Mitra, A.; Martinez-Rosas, M.; Bradley, A.L. Room-temperature ultraviolet luminescence from γ-CuCl grown on near lattice-matched silicon. *J. Appl. Phys.* **2005**, *98*, 113512. [CrossRef]
14. Zhou, Y.Y.; Lu, M.K.; Zhou, G.J.; Wang, S.F. Preparation and photoluminescence of gamma-CuI nanoparticles. *Mater. Lett.* **2006**, *60*, 2184–2186. [CrossRef]
15. Kurisu, H.; Nagoya, K.; Nakayama, N.; Yamamoto, S.; Matsura, M. Exciton and biexciton properties of CuCl microcrystals in an SiO_2 matrix prepared by sputtering method. *J. Lumin.* **2000**, *87–89*, 390–392. [CrossRef]
16. Fukumi, K.; Chayahara, A.; Kageyama, H.; Kadono, K.; Akai, T.; Mizoguchi, H.; Horino, Y.; Makihara, M.; Fujii, K.; Hayakawa, J. Formation process of CuCl nano-particles in silica glass by ion implantation. *J. Non-Cryst. Sol.* **1999**, *259*, 93–99. [CrossRef]
17. Yingjie, Z.; Yitai, Q.; Youfou, C. γ-Radiation synthesis of nanocrystalline powders of copper (I) halides. *Mater. Sci. Eng. B* **1999**, *57*, 247–250. [CrossRef]
18. Natarajan, G.; Maydannik, P.S.; Cameron, D.C.; Akopyan, I.; Novikov, B.V. Atomic layer deposition of CuCl nanoparticles. *Appl. Phys. Lett.* **2010**, *97*, 241905. [CrossRef]
19. Maydannik, P.S.; Natarajan, G.; Cameron, D.C. Atomic layer deposition of nanocrystallite arrays of copper(I) chloride for optoelectronic structures. *J. Mater. Sci.-Mater. Electron.* **2017**, *28*, 11695–11701. [CrossRef]
20. Lucas, O.; O'Reilly, L.; Natarajan, G.; McNally, P.J.; Daniels, S.; Taylor, D.M.; William, S.; Cameron, D.C.; Bradley, A.L.; Mitra, A. Encapsulation of the heteroepitaxial growth of wide bandgap-CuCl on silicon substrates. *J. Cryst. Growth* **2006**, *287*, 112–117. [CrossRef]

21. George, S.M.; Ott, A.W.; Klaus, J.W. Surface chemistry for atomic layer growth. *J. Phys. Chem.* **1996**, *100*, 13121–13131. [CrossRef]
22. Cameron, D.C.; Krumpolec, R.; Ivanova, T.V.; Homola, T.; Černák, M. Nucleation and initial growth of atomic layer deposited titanium oxide determined by spectroscopic ellipsometry and the effect of pretreatment by surface barrier discharge. *Appl. Surf. Sci.* **2015**, *345*, 216–222. [CrossRef]
23. Humlíček, J.; Šik, J. Optical functions of silicon from reflectance and ellipsometry on silicon-on-insulator and homoepitaxial samples. *J. Appl. Phys.* **2015**, *118*, 195706. [CrossRef]
24. Sesselmann, W.; Chuang, T.J. The interaction of chlorine with copper: I. Adsorption and surface reaction. *Surf. Sci.* **1993**, *176*, 32–66. [CrossRef]
25. Vasquez, R.P. CuCl by XPS. *Surf. Sci. Spect.* **1993**, *2*, 138–143. [CrossRef]
26. Vasquez, R.P. $CuCl_2$ by XPS. *Surf. Sci. Spect.* **1993**, *2*, 160–164. [CrossRef]
27. Martin-Vosshage, D.; Chowdari, B.V.R. XPS Studies on $(PEO)_nLiClO_4$ and $(PEO)_nCu(ClO_4)_2$ polymer electrolytes. *J. Electrochem. Soc.* **1995**, *142*, 1442–1446. [CrossRef]
28. Wren, A.G.; Phillips, R.W.; Tolentino, C.U. Surface-reactions of chlorine molecules and atoms with water and sulfuric-acid at low-temperatures. *J. Colloid Interface Sci.* **1979**, *70*, 544–557. [CrossRef]
29. Yamamoto, Y.; Konno, H. Ylide-metal complexes. X. An X-ray photoelectron spectroscopic study of triphenylmethylenephosphorane and gold- and copper-phosphorane complexes. *Bull. Chem. Soc. Jpn.* **1986**, *59*, 1327–1330. [CrossRef]
30. Beamson, G.; Briggs, D. *High Resolution XPS of Organic Polymers: The Scienta ESCA300 Database*; Wiley: New York, NY, USA, 1992.
31. Klein, J.C.; Proctor, A.; Hercules, D.M.; Black, J.F. X-ray excited Auger intensity ratios for differentiating copper-compounds. *Anal. Chem.* **1983**, *55*, 2055–2059. [CrossRef]
32. Wagner, C.D. Chemical-shifts of auger lines, and Auger parameter. *Faraday Discuss. Chem. Soc.* **1975**, *60*, 291–300. [CrossRef]
33. Ressan, M.; Furlani, C.; Polzonetti, G. XPS of coordination-compounds—Additive ligand effect in some copper(I) and copper(II) chelates with 1,2 phosphino (or phosphine oxide)-sulfido ethane ligands. *Polyhedron* **1983**, *2*, 523–528. [CrossRef]
34. Parmigiani, F.; Depero, L.E.; Minerva, T.; Torrance, J.B. The fine-structure of the Cu $2p_{3/2}$ X-ray photoelectron-spectra of copper-oxide based compounds. *J. Electron. Spectrosc. Relat. Phenom.* **1992**, *58*, 315–323. [CrossRef]
35. Kishi, K.; Ikeda, S. X-ray photoelectron spectroscopic study of the reaction of evaporated metal films with chlorine gas. *J. Phys. Chem.* **1974**, *78*, 107–112. [CrossRef]
36. Li, X.; Liu, Z.; Kim, J.; Lee, J-Y. Heterogeneous catalytic reaction of elemental mercury vapor over cupric chloride for mercury emissions control. *Appl. Catal. B Environ.* **2013**, *132–133*, 401–407. [CrossRef]
37. Biesinger, M.C.; Lau, L.W.M.; Gerson, A.R.; Smart, R.S.C. Resolving surface chemical states in XPS analysis of first row transition metals, oxides and hydroxides: Sc, Ti, V, Cu and Zn. *Appl. Surf. Sci.* **2010**, *257*, 887–898. [CrossRef]
38. Van der Laan, G.; Westra, C.; Haas, C.; Sawatzky, G.A. Satellite structure in photoelectron and auger-spectra of copper dihalides. *Phys. Rev. B* **1981**, *23*, 4369. [CrossRef]
39. Finster, J.; Klinkenberg, E.-D.; Heeg, J.; Braun, W. ESCA and SEXAFS investigations of insulating materials for ULSI microelectronics. *Vacuum* **1990**, *41*, 1586–1589. [CrossRef]
40. O'Reilly, L.; Natarajan, G.; McNally, P.J.; Cameron, D.; Lucas, O.F.; Martinez-Rosas, M.; Bradley, L.; Reader, A.; Daniels, S. Growth and characterisation of wide-bandgap, I–VII optoelectronic materials on silicon. *J. Mater. Sci. Mater. Electron.* **2005**, *16*, 415–419. [CrossRef]
41. Natarajan, G.; Mitra, A.; O'Reilly, L.; Daniels, S.; Cameron, D.C.; McNally, P.J.; Lucas, O.F.; Bradley, L. Optical investigations on sputtered CuCl thin films. In *Progress in Semiconductor Materials V: Novel Materials and Electronic and Optoelectronic Applications, Proceedings of the MRS Symposium, Boston, MA, USA, 28 November–1 December 2005*; Olafsen, L.J., Biefeld, R.M., Wanke, M.C., Saxler, A.M., Eds.; Cambridge University Press: New York, NY, USA, 2006; pp. 151–156.
42. Natarajan, G.; Mitra, A.; Daniels, S.; Cameron, D.C.; McNally, P.J. Temperature dependent optical properties of UV emitting γ-CuCl thin films. *Thin Solid Films* **2008**, *516*, 1439–1442. [CrossRef]
43. Certier, M.; Wecker, C.; Nikitine, S. Zeeman effect of free and bound excitons in CuCl. *J. Phys. Chem. Sol.* **1969**, *30*, 2135–2142. [CrossRef]

44. Cardona, M. Optical properties of the silver and cuprous halides. *Phys. Rev.* **1963**, *129*, 69–78. [CrossRef]
45. Akopyan, I.K.; Golubkov, V.V.; Dyatovla, O.A.; Novikov, B.V.; Tsagan-Mandzhiev, A.N. Structure of copper halide nanocrystals in photochromic glasses. *Phys. Solid State* **2008**, *50*, 1352–1356. [CrossRef]
46. Carcia, P.F.; McLean, R.S.; Walls, D.J.; Reilly, M.H.; Wyre, J.P. Effect of early stage growth on moisture permeation of thin-film Al$_2$O$_3$ grown by atomic layer deposition on polymers. *J. Vac. Sci. Technol. A* **2013**, *31*, 061507. [CrossRef]

© 2018 by the authors. Licensee MDPI, Basel, Switzerland. This article is an open access article distributed under the terms and conditions of the Creative Commons Attribution (CC BY) license (http://creativecommons.org/licenses/by/4.0/).

Article

Catalytic Performance of Ag$_2$O and Ag Doped CeO$_2$ Prepared by Atomic Layer Deposition for Diesel Soot Oxidation

Tatiana V. Ivanova [1,*], Tomáš Homola [2], Anton Bryukvin [1] and David C. Cameron [2]

1. ASTRaL Team, Laboratory of Green Chemistry, School of Engineering Science, Lappeenranta University of Technology, Sammonkatu 12, FI-50130 Mikkeli, Finland; brykvin@gmail.com
2. R & D Centre for Low-Cost Plasma and Nanotechnology Surface Modification, Masaryk University, Kotlářská 267/2, 611 37 Brno, Czech Republic; tomas.homola@mail.muni.cz (T.H.); dccameron@mail.muni.cz (D.C.C.)
* Correspondence: tatiana.ivanova@student.lut.fi; Tel.: +358-401-978-567

Received: 31 May 2018; Accepted: 28 June 2018; Published: 4 July 2018

Abstract: The catalytic behaviour of Ag$_2$O and Ag doped CeO$_2$ thin films, deposited by atomic layer deposition (ALD), was investigated for diesel soot oxidation. The silver oxide was deposited from pulses of the organometallic precursor (hfac)Ag(PMe$_3$) and ozone at 200 °C with growth rate of 0.28 Å/cycle. Thickness, crystallinity, elemental composition, and morphology of the Ag$_2$O and Ag doped CeO$_2$ films deposited on Si (100) were characterized by ellipsometry, X-ray diffraction (XRD), X-ray photoelectron spectroscopy (XPS), atomic force microscopy (AFM), and field emission scanning electron microscopy (FESEM), respectively. The catalytic effect on diesel soot combustion of pure Ag$_2$O, CeO$_2$, and Ag doped CeO$_2$ films grown on stainless steel foil supports was measured with oxidation tests. Nominally CeO$_2$:Ag 10:1 doped CeO$_2$ films were most effective and oxidized 100% of soot at 390 °C, while the Ag$_2$O films were 100% effective at 410 °C. The doped films also showed much higher stability; their performance remained stable after five tests with only a 10% initial reduction in efficiency whereas the performance of the Ag$_2$O films reduced by 50% after the first test. It was concluded that the presence of Ag$^+$ sites on the catalyst is responsible for the high soot oxidation activity.

Keywords: silver oxide; cerium oxide; oxidation; diesel soot; catalysis; ALD

---

1. Introduction

The number of diesel-powered vehicles has increased rapidly in recent years due to their reduced fuel consumption and thus lower CO$_2$ emission compared to petrol engines. However, diesel engines produce a high content of nitrogen oxides (NO$_x$) and particulate matter (PM) in their exhaust [1]. These emissions have a negative impact on human health causing respiratory, cardiovascular, and lung diseases, as well as on the environment such as disruption of the natural growth of plants and pollution of air, water, and soil [2,3]. Even though it is likely that many diesel engines will be replaced by petrol or electric engines in the future, there will still be a great need for diesel exhaust cleaning for some time to come.

In order to remove soot from the exhaust, diesel particulate filters (DPFs) are widely used [1]. Conventional DPFs require periodic regeneration by increasing the temperature of the exhaust gases to the soot combustion temperature, which is approximately 600 °C [4]. This method results in an increase in fuel consumption and clogging of the DPF by ash resulting in a slow increase of back pressure in the exhaust [1].

The composition of the exhaust mixture also affects the catalytic activity. Oxygen and NO_2 are generally used to oxidize diesel soot. NO contained in the raw exhaust gas is oxidized with excess oxygen into NO_2. Therefore, the development of catalysts, which can produce highly reactive oxygen species from O_2 molecules and NO_2 from NO, is the key issue. The catalyst promotes NO to NO_2 oxidation and NO_2 is then transported via the gas phase over the soot particles, oxidizing carbon while being reduced back to NO [5,6].

The preferred solution for continuous regeneration of the DPF is a catalysed diesel particulate filter (C-DPF) [7]. The main requirements for the catalyst are a reduction in temperature at which soot combustion occurs and long-term thermal and chemical stability.

Ceria-based catalysts have been studied in depth for various environmental applications such as three-way catalysts (TWC) for automotive pollution control, fluid catalytic cracking (FCC), and fuel cells [8,9]. The high potential of ceria as a catalyst is due to its fast and reversible reduction to sub-stoichiometric phases (CeO_2–CeO_{2-x}) as well as the high mobility of oxygen ions in its crystal lattice [10]. However, the use of metal-doped oxide catalysts can improve the performance of the bare oxide due to the increased mobility of oxygen species or the facilitation of the redox mechanisms associated with oxygen release/adsorption [11].

The influence of doping elements on the catalytic properties of ceria has been reported on recently by many researchers. The catalytic activity of ceria can be enhanced by doping with isovalent (Ti^{4+}, Zr^{4+}, Hf^{4+}, Sn^{4+}, etc.) and aliovalent (Zn^{2+}, La^{3+}, Ag^+, Eu^{3+}, etc.) cations into the ceria lattice [12–15]. Furthermore, the beneficial influence of Rh, Pd, Cu, Au, and Ag supported CeO_2 catalysts has been reported on elsewhere [16–18]. These reports showed increased electron mobility between the cerium buffer layer and support, favouring the formation of oxygen vacancies in CeO_2. Rangaswamy et al. [19] studied CeO_2–Sm_2O_3 and CeO_2–La_2O_3 catalysts, which could oxidize 50% of diesel soot under loose contact mode at 517 and 579 °C, respectively.

Among the metal additives investigated so far, Ag-based materials are the most promising catalysts for oxidizing diesel soot at low temperatures. Aneggi et al. [20] reported the effect of Ag addition on various metal oxides (CeO_2, ZrO_2, Al_2O_3) during soot oxidation activity. They showed that Ag/CeO_2 and Ag/ZrO_2 catalysts have high soot oxidation activity in the temperature region around 300 °C under tight contact mode. Haneda et al. [21] also performed isotopic transient kinetic analysis on Ag/ZrO_2 catalyst and concluded that the presence of Ag^+ sites in Ag/ZrO_2 was responsible for the high soot oxidation activity. Ag and Fe doped Mn_2O_3 catalysts were examined by Kuwahara et al. to enhance soot oxidation under tight contact mode and showed the T50 (the temperature for 50% of soot combustion) at 290 °C and at 328 °C, respectively. Based on their measurements, the mechanism of soot oxidation was proposed to be by the activated lattice oxygen species in Ag doped Mn_2O_3 catalyst via the redox of Ag^0/Ag_2O species [22].

Machida et al. [9] investigated silver loading onto CeO_2 and showed the enhancement of catalytic activity for soot oxidation because of the enhanced generation of superoxide. Shimizu et al. [23] showed that the presence of Ag metal nanoparticles on CeO_2 significantly improved the reactivity of CeO_2 lattice oxygen during soot decomposition under oxygen and under inert atmosphere. In addition, a dopant of silver in CeO_2 may increase oxygen mobility due to a weakened Ce–O bond [24].

There are a number of methods of preparation of Ag doped CeO_2 catalysts such as co-precipitation [25], impregnation [19,26], and liquid-phase chemical reduction [20]. We chose to use atomic layer deposition (ALD). The benefits of ALD compared to the other methods are extreme film thickness uniformity, precise thickness control, excellent step coverage, and high reproducibility. The thickness of the films can be easily controlled by controlling the number of deposition cycles. Furthermore, the fact that ALD operates via self-limiting surface reactions in consecutive cycles means that doping materials can be introduced with greater control and tuning than other deposition methods. ALD can be used to deposit catalytic coatings on high surface area porous powder supports or on geometrically complex structures [27] such as particulate filters in diesel engine exhaust systems.

In this study, we investigated the ALD of Ag_2O and Ag doped CeO_2 for catalytic applications in soot combustion under loose contact mode. The crystal structure, morphology, and composition properties of the deposited films were analysed. The effect of doping on the efficacy of soot combustion in annealing tests was also studied, paying particular attention to the doping concentration and oxidation state of silver in the CeO_2 thin films.

2. Materials and Methods

2.1. Catalyst Preparation

The deposition of Ag doped cerium oxide was carried out using an F-120 ALD reactor (ASM Microchemistry Ltd., Espoo, Finland). Thin films were deposited with different doping concentrations at a reaction temperature of 200 °C. The cyclic nature of ALD means that pulses of dopant can easily be incorporated into the main process. The desired composition of catalytic thin film can be achieved by depositing n cycles of the base CeO_2 material (where n can be varied to suit the required doping level) with one cycle of the doping material inserted (Figure 1). The supercycle (n + 1), which consists of two individual ALD processes, was repeated x times until the required film thickness was achieved. The process for CeO-based material contains two half-cycles using 2,2,6,6-tetramethyl-3,5-heptadionatecerium $Ce(C_{11}H_{19}O_2)_4$ ($Ce(thd)_4$ for brevity) and O_3 as precursors. The Ag doping material also comprises two half-cycles of Ag precursor (see below) and O_3.

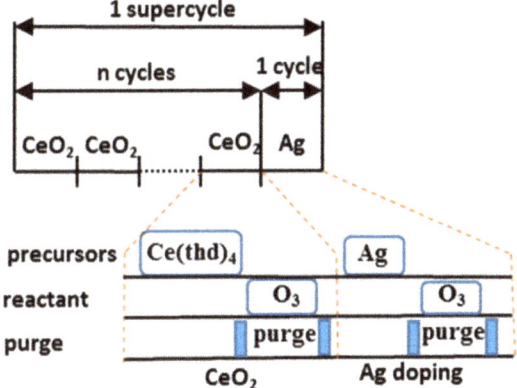

Figure 1. A schematic representation of the atomic layer deposition (ALD) supercycle used to deposit Ag doped CeO_2 catalytic thin films.

$Ce(thd)_4$ (Volatec, Porvoo, Finland) and trimethylphosphine (hexafluoroacetylacetonalo)-silver $Ag(CF_3COCHCOCF_3)P(CH_3)_3$ (($hfac)Ag(PMe_3)$, 99%; Strem Chemicals, Newburyport, MA, USA) were used as Ce and Ag precursors respectively. Ozone O_3 was used as the co-reactant in both cases and was generated by an ozone generator (Wedeco Modular 4HC Lab, Herford, Germany) from a pure oxygen (>99.999%) source. Ozone concentration was 120 g/m^3. Nitrogen (>99.999%) was used as a carrier and purge gas between precursor pulses. The pressure in the reactor was approximately 1 mbar. $Ce(thd)_4$ and $(hfac)Ag(PMe_3)$ were evaporated at 160 and 80 °C, respectively to achieve sufficient vapour pressure. The saturated deposition rate in the ALD supercycle should be obtained when the two individual ALD processes are in saturation. We used the previously optimized CeO_2 ALD process parameters: 1.5 s $Ce(thd)_4$ dose, 2.5 s purge, 2.5 s O_3 dose, 2.5 s purge [28]. The pulse time for $(hfac)Ag(PMe_3)$ was varied from 0.5 to 4 s in 0.5 s steps while keeping the O_3 pulse time constant at 2.5 s. After finding the optimal pulse time for $(hfac)Ag(PMe_3)$ the pulse time for O_3 was determined with the same method with 2.5 s of the purge time.

In order to achieve doping of CeO_2 with Ag, one supercycle consisted of n CeO_2 cycles, with n equal to 10, 20, and 30, and 1 cycle of Ag. The supercycle was repeated 150, 75, and 50 times for CeO_2:Ag ratios 10:1, 20:1, and 30:1 respectively in order to achieve comparable film thicknesses.

Silicon substrates <100> (Si-Mat, Kaufering, Germany) were used for process development while stainless steel foil AISI 316 with a thickness of 0.025 mm (Goodfellow Cambridge Ltd., London, UK) was used as a substrate for soot burning tests. Stainless steel foil was chosen because of its relatively low weight, which reduced the error during weighing of samples to determine the amount of soot oxidation. The substrates were cut in pieces of 20 mm × 10 mm and cleaned using an ultrasonic bath with acetone, isopropanol, and deionized water consecutively, each with a time of 5 min and thereafter dried using compressed air.

2.2. Soot Deposition System

The soot deposition system was described in detail in the previous report [28]. Briefly, the diesel soot was generated with a Webasto diesel engine preheater (Webasto group, Stockdorf, Germany) from diesel fuel and air. The samples were placed on the heating plate and exposed to exhaust gases for 1.5 min. The amount of deposited soot was measured by weighing the samples.

2.3. Catalyst Characterization

A spectroscopic ellipsometer J.A. Woollam M-2000UI (J.A. Woollam Co., Lincoln, NE, USA) was used to determine the catalytic film thickness. This was obtained by using a Cauchy model for fitting.

The surface topography of the catalytic thin films was evaluated with field emission scanning electron microscopy (FESEM) Hitachi S-4800 (Hitachi, Tokyo, Japan). An atomic force microscopy (AFM) (Park NX10, Park Systems, Suwon, Korea) was utilized for analysing the film morphological properties such as roughness and cluster size. All measurements were done in non-contact mode with a cantilever force constant of 42 N/m.

The crystal structure of the catalytic materials was studied by a Rigaku SmartLab® Type F XRD (Rigaku, Tokyo, Japan; Cu-Kα radiation, λ = 1.5418 Å). The grazing incidence X-ray diffraction (GIXRD) scan was collected with a grazing incidence angle of 0.5°. Scan speed of 0.9°/min and 2θ values from 20° to 90° were used. The high resolution scan was taken with speed 0.045°/min.

The surface chemical composition, bonding properties and analysis of impurities in the deposited films were investigated by X-ray photoelectron spectroscopy (XPS) using an ESCALAB 250Xi (Thermo Scientific, Loughborough, UK) with a monochromated Al-Kα (energy of 1486.7 eV) X-ray source in the constant pass energy mode with a value of 50 eV. For high resolution spectra of Ag 3d a pass energy of 20 eV with resolution 0.1 eV was used. Charging compensation by an electron flood source was used in all measurements to minimize binding energy shifts. The binding energy of C 1s was set to be 284.5 eV as an internal standard for calibration. Sputtering by Ar$^+$ ions at 2 kV for 20 s was applied to remove surface contaminations and obtain actual carbon levels in the films. Deconvolution and fitting of the obtained peaks were made with Avantage software (Version 5.938) using Smart type background and applying 90:10 Gaussian-Lorentzian peaks.

2.4. Catalytic Activity

Five samples of Ag doped CeO_2, pure CeO_2 and pure Ag_2O catalytic coatings were separately deposited on stainless steel foils for evaluation of catalytic activity. For measurement of the amount of combusted soot, annealing tests were performed by placing the catalytic samples covered with soot into an oven and measuring the weight loss of the samples over 2 h in the temperature range of 300–490 °C in ambient atmosphere. The molar ratio of catalysts to carbon was approximately 1:80 with average soot mass of 0.7 g. The annealing measurements were repeated 5 times to evaluate reproducibility of the prepared catalysts.

3. Results and Discussion

3.1. Ag_2O and Ag-Doped CeO_2 ALD Film Deposition

3.1.1. Ag_2O

The concept of the supercycle can be effective when two individual processes are compatible with each other and each is in saturation. Therefore, before the introduction of Ag as a dopant into the CeO_2 structure, silver oxide thin films were deposited first on silicon substrates to find the self-limiting growth regime of the ALD process. This could be assessed by measuring the growth rate as a function of the amount of precursor delivered into the reactor. The ALD saturation growth study was carried out at deposition temperature of 200 °C. This temperature was chosen to match the reaction temperature of CeO_2 and it has been shown that the nucleation period of Ag is significantly shortened at this temperature [29]. Figure 2 shows the saturation curves for (hfac)Ag(PMe$_3$) and O_3. The pulsing time of (hfac)Ag(PMe$_3$) or O_3 was adjusted within the range of 0.5–4 s while the other precursor pulse was fixed at 2.5 s. The saturation of (hfac)Ag(PMe$_3$) and O_3 precursors can be seen at 2.5 s with constant Ag_2O growth rate of 0.28 Å/cycle. No further increase of the growth rate was noticed after increasing the precursor pulse time up to 4 s.

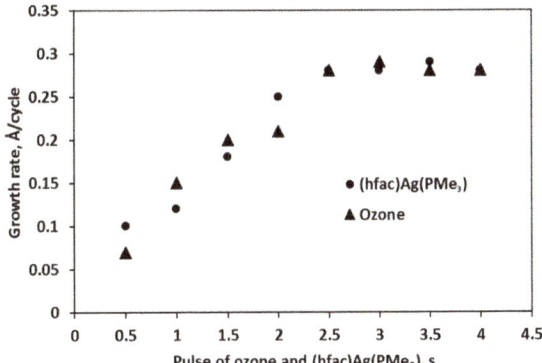

Figure 2. The effect of ozone and (hfac)Ag(PMe$_3$) pulse time on the growth rate of silver oxide at 200 °C reactor temperature.

Figure 3 shows the Ag_2O film thickness versus the number of ALD cycles from 10 to 750 deposited at 200 °C. From Figure 3 it can be seen that the thickness increases linearly after 100 ALD cycles. The incorporation of silver atoms is directly related to the density of hydroxyl groups on the substrate surface that act as adsorption sites for (hfac)Ag(PMe$_3$) molecules. As illustrated in Figure 3, the initial growth rate of the films per cycle (GPC) is substrate dependent at the start of the ALD process and it takes about 100 cycles to obtain a stable GPC of 0.28 Å/cycle. The film growth can be separated into two regimes: an island-like growth for the first 100 cycles and layer-by-layer growth as is expected from the proceeding atomic layer deposition. If the bare Si has a higher density of reactive sites compared to the deposited Ag_2O or these sites have a higher reactivity than the reactive sites on Ag_2O, then the growth rate will initially be higher until a complete film is formed. Alternatively, some hfac ligands may not be completely removed by ozone and may remain bound to the surface. These comments are expanded on in Section 3.1.2.

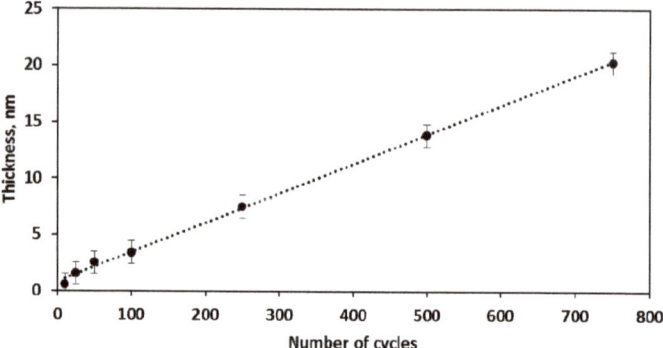

Figure 3. Thickness of the silver oxide film as a function of the number of deposition cycles at the deposition temperature of 200 °C.

3.1.2. Ag Doped CeO_2

The film thickness after deposition with different CeO_2:Ag ratios was measured by spectroscopic ellipsometry. The total number of CeO_2 cycles was chosen to be 1500 while the number of Ag cycles was varied from 50 up to 150 according to the supercycle configuration. Growth rate of the films per cycle (GPC) over the total number of ALD cycles is shown in Figure 4 as a function of Ag dopant fraction in CeO_2 film. The reduction of GPC with increasing Ag dose observed in Figure 4 could be the result of a slight etching of CeO_2 by (hfac)Ag(PMe$_3$), but it could also be due to nucleation delay and the inhibition of CeO_2 growth after (hfac)Ag(PMe$_3$) pulsing.

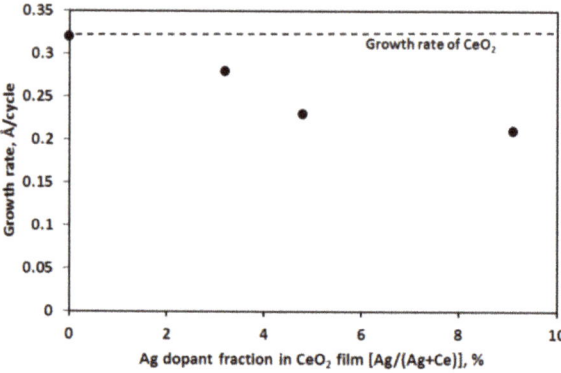

Figure 4. The effect of Ag concentration on the growth rate of Ag doped CeO_2 thin films at 200 °C reactor temperature.

The comparison of the two calculated and experimental silver doping concentrations inside the Ag doped CeO_2 films is shown in Figure 5. The calculated value is determined by the ratio of the number of doping ALD cycles divided by the total number of ALD cycles in one supercycle ($1/(n + 1)$). The experimental value is obtained from XPS measurements. It is worth emphasizing that the concentrations of dopant do not relate directly to the doping efficiency. Some of the dopant silver atoms might have formed silver oxide or alloy clusters rather than only doping the film, as will be discussed in Section 3.2.

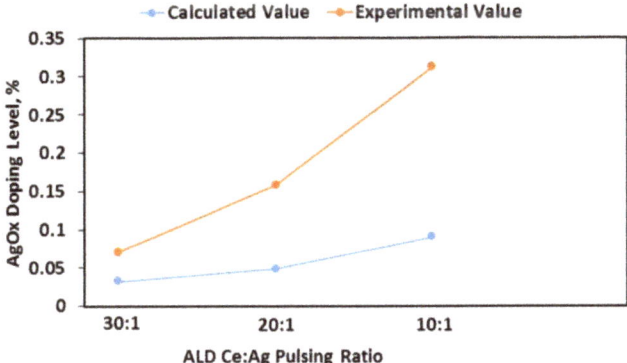

Figure 5. Relationship between CeO$_2$:Ag pulsing ratio vs. silver molar percent in the corresponding Ag doped CeO$_2$ thin films.

The hypothetical deposition of Ag doped CeO$_2$ by ALD can be explained as follows: (a) Ce(thd)$_4$ adsorbs on nucleation sites (–OH) and dissociates into the attached fraction Ce(thd)$_x$*, where * designates surface species, with x varying from 1 to 3 depending on the number of OH sites it bonds to. Most probably, some –OH groups remain unreacted, due to steric hindrance; (b) Ozone regenerates O* groups which can act as nucleation sites during subsequent (hfac)Ag(PMe$_3$) exposures. The O$_3$ half-cycle also probably results in the formation of OH groups because of decomposition of the precursor ligand; (c) (hfac)Ag(PMe$_3$) may adsorb on Ce–O* or Ce–OH nucleation sites and on unreacted –OH groups, and dissociates into Ce(hfac)* and Ag(hfac)* species; (d) ozone may react with Ce(hfac)* and Ag(hfac)* species regenerating O* groups for further Ce(thd)$_4$ exposures. We propose that during Ce(thd)$_4$ treatments, not all –OH or regenerated O* groups can react with the precursor or, as mentioned above, hfac ligands remain bound to the surface and cannot be completely removed by ozone. This statement is supported by XPS measurements, where a high level of impurities was noticed, and is discussed in detail in Section 3.2. It could be the reason for formation of Ag$_2$O clusters in the films with higher concentration of Ag doping, as will be considered during AFM and SEM analysis. With more CeO$_2$ ALD cycles, more nucleation sites are generated and this facilitates the growth of CeO$_2$ thin film with Ag as a dopant.

3.2. Catalyst Characterization

Ag$_2$O and Ag-Doped CeO$_2$

In examining the nucleation effect and morphology of the resulting ALD of Ag$_2$O, several AFM images were taken with variable ALD cycles (Figure 6). The average size of Ag$_2$O nanoparticles after 25 ALD cycles was around 22 nm with film roughness of 1.5 nm. After 100 ALD cycles, large nanoparticles with size of around 40 nm were present with a high nanoparticle density (Figure 6a) and overall surface coverage. The AFM image of 500 cycles of Ag$_2$O film is shown in Figure 6b, which demonstrates that the surface is now fully covered with silver, having grain sizes between 40 and 46 nm. As was shown in Figure 3, the nucleation region for Ag$_2$O ALD occurs over 100 cycles; Ag$_2$O films can nucleate and grow by the Volmer–Weber (VW) growth mechanism, where the deposited atoms form islands or clusters and three dimensional aggregates on the substrate. Growth of these clusters, along with coarsening, can be a cause of rough thin films on the substrate surface [30].

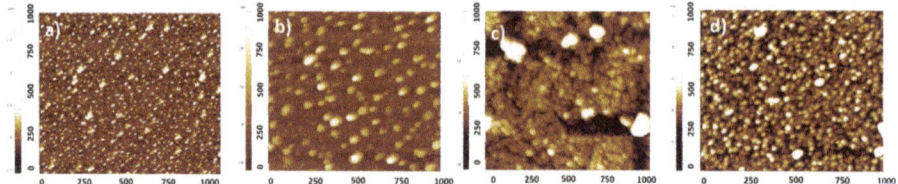

Figure 6. The atomic force microscopy (AFM) non-contact mode images of Ag_2O deposited with: (**a**) 25 cycles (R_a 0.7 nm); (**b**) 100 cycles (R_a 1.5 nm); (**c**) 250 cycles (R_a 1.6 nm); and (**d**) 500 cycles (R_a 1.7 nm). Axis scales are in nm.

All the Ag_2O thin films deposited on Si had a visual matt finish, which is a sign of rough microstructure. SEM studies supported the AFM results (Figure 7). The 250 and 500 ALD cycles films were confirmed to consist of particles with widely different sizes as a result of coalescence and secondary nucleation on existing particles.

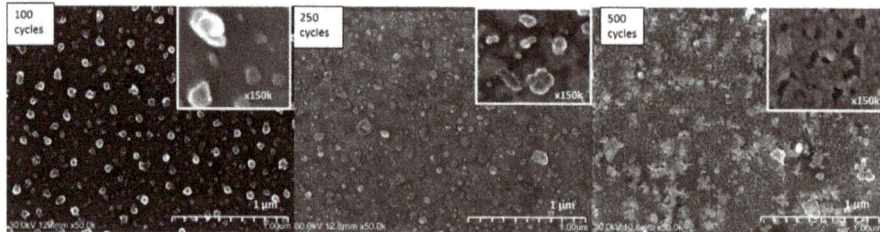

Figure 7. Scanning electron microscope (SEM) images of Ag_2O thin films deposited with different number of cycles at 200 °C resulting in different thicknesses (5.8, 8.8, and 13.8 nm).

The AFM and SEM analyses on Figures 8 and 9 show that the surface morphology of Ag doped CeO_2 films changes in accordance with silver concentration in the film. It can be seen that the reduction of Ag concentration dramatically decreases crystal and cluster sizes in doped films. Figures 8a and 9a suggest that higher concentration of Ag (CeO_2:Ag 10:1) inhibits CeO_2 growth so that the Ag nuclei are not covered with CeO_2 and so the next Ag cycle nucleates more easily on top of the Ag and can therefore form bigger crystals of about 50 nm in size. Such large nanoparticles were noticed on pure Ag_2O and described above. With lower doping concentration of Ag in CeO_2 thin films, no crystals larger than 25 nm were noticed.

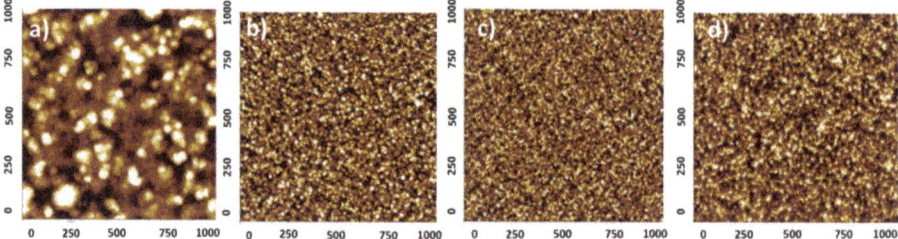

Figure 8. The AFM non-contact mode images of Ag doped CeO_2 in different CeO_2:Ag ratios deposited at 200 °C: (**a**) 10:1; (**b**) 20:1; (**c**) 30:1; and (**d**) CeO_2. The scales are in nm.

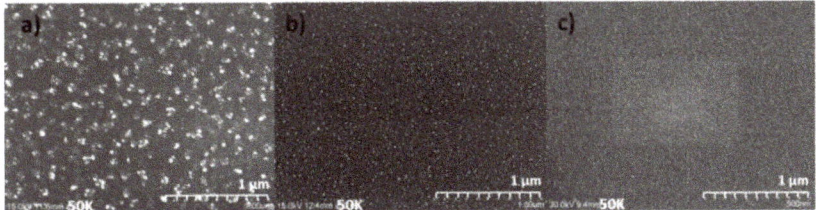

Figure 9. SEM of Ag doped CeO_2 in different CeO_2:Ag ratios (**a**) 10:1, (**b**) 20:1, and (**c**) 30:1 deposited at 200 °C.

Figure 10a shows XRD spectra of Ag_2O, CeO_2, and Ce:Ag 10:1, 20:1, and 30:1 Ag doped CeO_2 films in the 2θ range of 20°–90°. The Ag_2O films showed strong X-ray diffraction peaks at 2θ = 32.7° and 38.3° related to the (111) and to (200) cubic planes of Ag_2O, respectively (ICCD Card No: 00-41-1104). They revealed that the films deposited at 200 °C contained only Ag_2O crystallites. The diffraction peaks at 2θ angles of 28.6°, 33.6°, 47.6°, and 56.3° can be identified for all the other samples and attributed to (111), (200), (220), and (311) planes of cubic cerium oxide, respectively (COD database, card No 9009008). The XRD patterns of CeO_2, Ag_2O, and Ag doped CeO_2 in the 2θ range 24–40° were expanded (Figure 10b) to analyse the position of CeO_2 and Ag_2O reflections in the X-ray spectra in more detail. Figure 10b shows that the cerium oxide peak intensity and the shape of (200) plane reflection in the Ag doped CeO_2 samples decreased and broadened, respectively, with increasing Ag doping, compared to the pure CeO_2 catalyst.

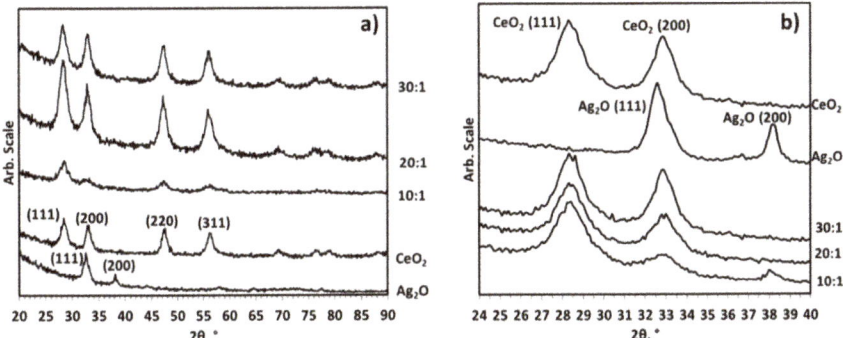

Figure 10. (**a**) X-ray diffraction (XRD) patterns of Ag_2O, CeO_2 and Ag doped CeO_2 with different Ce:Ag ratios 10:1, 20:1, and 30:1 deposited at 200 °C. (**b**) XRD patterns of slow scans of CeO_2, Ag_2O and Ag doped CeO_2 with different Ce:Ag ratios 10:1, 20:1, 30:1 in the 2θ range of 24–40°.

The mean grain size, assuming spherical grains, of CeO_2 can be determined from the full width at half maximum (FWHM) of the (111) XRD peak, through Scherrer's equation. The grain sizes calculated from the (111) plane reflection of CeO_2 are indicated in Table 1. XRD analyses confirm the decrease of the CeO_2 crystallite size of (111) plane as the Ag amount increases. This behaviour is related to the occurrence of lattice defects due to the presence of the dopant, which leads to deformations in the crystalline structure and smaller crystallites. The XRD data do not show any peaks related to Ag species for CeO_2:Ag 20:1 and 30:1 catalysts, which is, most probably, demonstrative of a high distribution of the dopants in the CeO_2 samples. The CeO_2 (200) and Ag_2O (111) appear at very similar positions. However, for higher amount of Ag doping Ce:Ag 10:1, a small peak from the Ag_2O (200) plane reflection can be observed (Figure 10b). In the CeO_2 lattice, the radius of Ce^{4+} ion is 0.97 Å. However, the ionic radius of Ag^+ ions is 1.28 Å [31]. As such, substitution or replacement of Ag^+ ions for Ce^{4+} ions in the CeO_2 lattice requires high energy [32] and from the XRD there is no evidence of significant substitutional doping during ALD of Ag doped CeO_2 thin films from a shift in the position of the CeO_2 peaks to smaller angles. Ag_2O forms as metal oxide or alloy clusters in CeO_2:Ag 10:1 catalyst and inhibits CeO_2 crystal formation.

Table 1. Grain size of CeO_2 (111) plane reflection based on Scherrer's equation.

Catalyst	Grain Size of CeO_2 (111) (nm)
CeO_2	10.2
Ce:Ag 30:1	8.7
Ce:Ag 20:1	7.1
Ce:Ag 10:1	6.2

XPS was employed to analyse the chemical state of the as-deposited Ag_2O thin films and the Ag doping in CeO_2 films, which were controlled by varying the Ce:Ag supercycle binary process pulse ratio. The information on silver and cerium oxidation states was obtained from the high resolution Ag 3d and Ce 3d spectra after Ar^+ bombardment to exclude surface contaminations.

Using the values of surface atomic composition from Table 2, an estimation of the O/Ce and O/Ag atomic ratio can be obtained. The ratio O/Ce for the cerium oxide deposited at 200 °C is around two, which indicates that the pure CeO_2 is stoichiometric. The carbon impurity level is around 21 at.%, which arise from the $Ce(thd)_4$. The Ag to O ratio in the Ag_2O films was estimated to be close to 2:1, which indicates that the film primarily consists of Ag_2O with 14.2% of carbon, 0.5% of F, and 3% of N as the main impurities in that film. With regard to the Ag/Ce surface atomic ratio, an important enhancement with Ag loading is observed, indicative of an increase in the number of Ag surface atoms. We found from survey spectra that by changing the Ce:Ag ratio from 30:1 to 10:1 the amount of Ag increases from ~2 at.% to ~9.7 at.%., as measured by XPS (Table 2).

Table 2. Surface elemental composition of Ag doped CeO_2, Ag_2O, and CeO_2 thin films.

Catalyst	Surface Composition (at.%)						Ag/Ce (from Survey)
	Ce 3d	Ag 3d	O 1s	C 1s	F 1s	N 1s	
Ce:Ag 10:1	21.4	9.7	46.1	13.9	8.1	6	0.453
Ce:Ag 20:1	23.9	4.5	43.8	15.7	7.1	5	0.188
Ce:Ag 30:1	26.6	2.0	39.9	17.4	5.7	3.6	0.075
Ag_2O	–	56	27.3	14.2	0.5	3	–
CeO_2	25.9	–	52.9	21.5	–	–	–

High resolution spectra of the Ag 3d peaks of Ag_2O and Ag doped CeO_2 thin films with the nominal ratio Ce:Ag from 30:1 to 10:1 give us indications of the chemical state of Ag atoms (Figure 11). The pure Ag_2O films showed only one peak at 368.2 eV. The binding energy which has been observed for pure Ag_2O thin film is 367.2 eV, which consists of the dominant oxidation state Ag^+ [33]. The spectrum here shows a shift of ~1 eV in the peak position compared to the previously found results which may be due to sample charging. The spectrum showed core level binding energies at about 368.2 ± 0.1 eV and 374.2 ± 0.1 eV related to the Ag $3d_{5/2}$ and Ag $3d_{3/2}$ respectively with spin orbit separation of 6 eV [34]. Each Ag 3d level in Ag doped CeO_2 films can be deconvoluted into three peaks, with corresponding binding energies 368.2, 369.2, and 367.2 eV, which are consistent with those of Ag^+, Ag^0, and Ag^{2+} (Table 3), allowing for the shift due to sample charging [35–37]. The estimated percentages of the three peaks, shown in Table 3, indicates that with increasing Ag doping concentration in CeO_2 films from 30:1 to 10:1, the Ag^+ oxidation state also grows from 38.4% to 85%, respectively. At the same time the Ag^0 oxidation state decreases from 59% to 9.7% for CeO_2 doped Ag films deposited with the ratio Ce:Ag from 30:1 to 10:1, respectively.

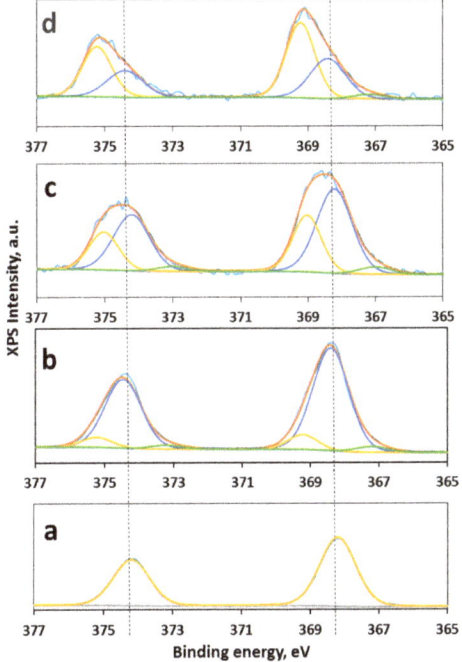

Figure 11. High resolution Ag 3d X-ray photoelectron spectroscopy (XPS) spectra of (**a**) pure Ag$_2$O and Ag doped CeO$_2$ in different CeO$_2$:Ag ratio (**b**) 10:1, (**c**) 20:1, and (**d**) 30:1.

Table 3. Binding energies and integrated peak areas of Ag 3d spin-orbit doublets in Ag$_2$O, Ag doped CeO$_2$ and CeO$_2$ thin films.

Catalyst	Compound (BE, eV)			Concentration of Ce^{4+} (at.%) (Excluding C)	Concentration of Ce^{3+} (at.%) (Excluding C)	Ce^{3+}/Ce^{4+} (%)
	Ag$^+$ (368.2 ± 0.1)	Ag^0 (369.2 ± 0.1)	Ag^{2+} (367.2)			
Ag$_2$O	100	–	–	–	–	–
Ce:Ag 10:1	85	9.7	4.5	77	23	29.8
Ce:Ag 20:1	59	35	6	80	20	25.0
Ce:Ag 30:1	38.4	59	2.6	82	18	21.9
CeO$_2$	–	–	–	83	17	20.4

Figure 11 and Table 3 show that at low doping concentration, Ag species mainly exist as Ag^0, while as doping concentration increases, Ag$^+$ species increase remarkably. It is likely that at low concentrations of Ag$_2$O doping, some of the silver oxide is reduced by the CeO$_2$; a similar effect has been reported on Ag$_2$O-doped TiO$_2$ [38]. For higher Ag dopant concentration, most of the silver present in the catalysts remains as cations and probably interacts with CeO$_2$ through the Ag–O bonds. Based on previous studies involving silver oxides [39–42], it can be concluded that some electrons may transfer from CeO$_2$ to the Ag dopant and there is strong interaction between the Ag species and the CeO$_2$ catalyst.

It is interesting to note that the concentration of Ce^{4+} decreased from 82% to 77% and the concentration of Ce^{3+} increased from 18% to 23% with increasing Ce:Ag doping from 30:1 to 10:1, respectively, further suggesting the existence of the interaction between Ag and CeO$_2$. This is probably because the Ag$^+$ ions in Ag doped CeO$_2$ can partially substitute Ce^{4+} in the CeO$_2$ matrix in the form of Ce$_{1-x}$Ag$_x$O$_{2-\delta}$. As was shown earlier from the XRD spectra, there is no evidence of substitutional doping. However, the increasing Ag content also produced smaller crystallites so it may be that the increasing Ce^{3+} arises as a consequence of interaction between the CeO$_2$ and the Ag in the disordered regions at the grain boundaries. In summary, Ag atoms deposited on a stoichiometric CeO$_2$ surface tend to result in reduction of the Ce ions, which leads to the stabilization of the Ag in the +1 oxidation state. These results are in good agreement with the literature reports [43–45].

3.3. Catalytic Activity of Ag$_2$O, CeO$_2$, and Ag-doped CeO$_2$ catalysts

The evaluation of the catalytic activity of Ag$_2$O, CeO$_2$, and Ag doped CeO$_2$ thin films deposited at 200 °C on stainless steel foil was carried out to show the effectiveness of the catalysts for soot combustion applications. The annealing tests of carbon soot, which was generated from diesel fuel, was carried out under ambient air environment inside an oven in the temperature range 300–490 °C for 2 h.

Figure 12 shows histograms of soot conversion vs. annealing temperature for catalytic and non-catalytic combustion. The conversion is defined as:

$$C(\%) = \frac{M_0 - M}{M_0} \times 100\% \quad (1)$$

where is M_0 is the initial soot mass, and M is the amount of soot left on the catalyst after burning by heating up to a given T value. Weight loss values were obtained by weighing the samples before and after the annealing test which continued for 2 h.

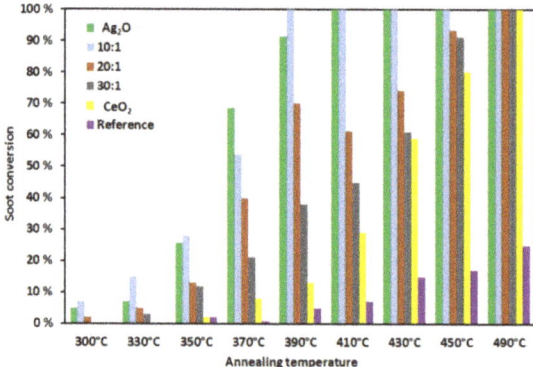

Figure 12. Conversion of oxidized soot on silver oxide, cerium oxide, and silver doped cerium oxide in ratio CeO$_2$:Ag 10:1, 20:1, 30:1 thin films deposited on stainless steel foil at 200 °C vs. annealing temperature over 2 h. The measurement uncertainty is approximately 5%.

The conversion of oxidized soot was demonstrated on non-catalysed reference steel foil, and on cerium oxide, silver oxide, and Ag doped CeO$_2$ thin films deposited on stainless steel foil at 200 °C in loose contact mode (Figure 12). As expected, complete soot conversion on the uncoated reference sample was only achieved at 600 °C. All the catalysts were effective in promoting combustion at temperatures below 490 °C. Although the soot was well oxidized on pure CeO$_2$ and Ag$_2$O thin films themselves, Ag loading into CeO$_2$ thin films caused a significant enhancement of soot oxidation rate, in accordance with previous reports [28]. It is noteworthy that the soot oxidation activity of Ag doped CeO$_2$ was different depending on the dopant concentration. The sample having the maximum silver loading CeO$_2$:Ag 10:1 and pure Ag$_2$O showed the lowest oxidation temperature of 300 °C and complete combustion of the soot was achieved below 390 and 410 °C under real-world loose contact conditions, respectively. Table 4 lists the characteristic temperatures, where the ignition temperature (T_i) is the temperature at which the combustion began, and the final temperature (T_f) is the temperature at which the soot was completely oxidized.

Table 4. Catalytic performance for soot oxidation.

Catalyst	T_i (°C)	T_f (°C)
None	410	600
CeO$_2$	350	490
CeO$_2$:Ag 10:1	300	390
CeO$_2$:Ag 20:1	300	490
CeO$_2$:Ag 30:1	330	490
Ag$_2$O	300	410

The catalysts with silver doping concentration of Ce$_2$O:Ag 20:1 and 30:1 showed lower performance for 100% soot oxidation at T_f = 490 °C. This indicates that the highest concentration of Ag$^+$ ions, which is contained mostly in CeO$_2$:Ag 10:1 catalyst, can effectively promote 100% soot oxidation at T_f = 390 °C due to the oxygen species formed on Ag$^+$ sites. Zou et al. [46] proposed that during silver oxide decomposition, its released oxygen migrates to soot surfaces to form carbon–oxygen intermediates, which are subsequently oxidized further. Finally, the adsorbed oxygen on the silver promotes the regeneration process. The redox reaction of Ag$^+$ and the active oxygen species prevail in the reaction at low combustion temperatures. In addition, the silver oxide contributes to the adsorption of reactants to form complex π bonds that are significant for the formation of peroxide and superoxide species [47]. It is worth mentioning that higher concentration of Ce^{3+} contained in CeO$_2$:Ag 10:1 catalyst can also form more active oxygen vacancies, which promote the activation of adsorbed oxygen to form superoxides in the lattice. These types of oxygen react with soot efficiently [48].

Because the melting point of Ag$_2$O oxides is relatively low, compared with CeO$_2$ (2400 °C), the stability of Ag/CeO$_2$ catalysts during the soot oxidation is an important factor from the practical point of view. In order to gain information on the stability of Ag/CeO$_2$, repetitive activity tests were carried out in loose contact mode at annealing temperature 430 °C. Figure 13 shows the conversion of oxidized soot after five replicate trials, and the observed results showed acceptable reproducibility with relative standard deviation of less than 5%. The pure Ag$_2$O catalyst lost its catalytic activity immediately after the first trial from 100% to 50% of oxidized soot. Further use of Ag$_2$O catalyst (in the third, fourth, and fifth trials) led to significant restructuring of the film and total loss of catalytic properties. For this reason, bulk Ag$_2$O cannot be used in catalytic systems operating at higher temperature (above 300 °C).

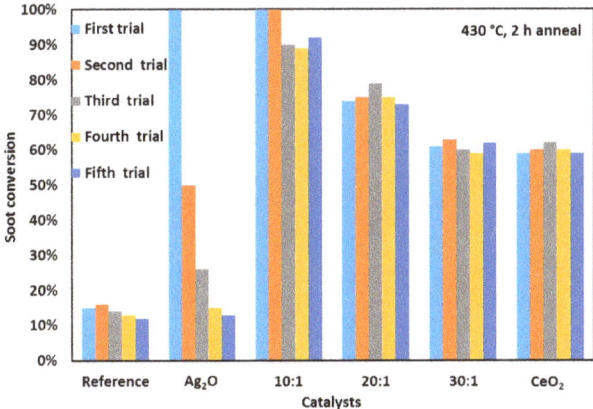

Figure 13. Repetitive soot oxidation in the presence of silver oxide, cerium oxide, and silver doped cerium oxide in ratio CeO$_2$:Ag 10:1, 20:1, and 30:1 thin films deposited on stainless steel foil at 200 °C vs. annealing temperature over 2 h at 430 °C.

It appears that CeO$_2$/Ag 10:1 lost 10% of its activity in the third combustion test, indicating the deactivation of Ag/CeO$_2$. However, during further trials this catalyst showed stable results of 90% oxidized soot. We can assume that Ag ions have a strong interaction with the CeO$_2$ catalyst and remain stable after durability tests. Other catalysts such as CeO$_2$:Ag 20:1, 30:1 and pure CeO$_2$ do not lose activity, since the combustion curves obtained from the first to the fifth tests were very similar.

4. Conclusions

We have demonstrated that ALD can be used to deposit Ag$_2$O and Ag doped CeO$_2$ catalysts at 200 °C with (hfac)Ag(PMe$_3$), Ce(thd)$_4$, and O$_3$ as precursors. The growth rate stabilized to a steady-state Ag$_2$O GPC of 0.28 Å/cycle after 100 ALD cycles. The silver doping concentration was finely tuned by setting the CeO$_2$:Ag cycle ratio to 10:1, 20:1, and 30:1. With increasing Ag concentration in the CeO$_2$ thin films, the overall GPC decreased from 0.32 down to 0.21 Å/cycles. AFM and SEM analyses showed that higher concentration of Ag (CeO$_2$:Ag 10:1) inhibited CeO$_2$ growth so that the Ag nuclei were not covered with CeO$_2$ and so the next Ag cycle nucleated more easily on top of the Ag and could therefore form bigger crystals. XRD and XPS analyses showed that the Ag$^+$ oxidation state dominated for CcO$_2$:Ag 10:1 catalysts and stoichiometric CeO$_2$ tended to be reduced from Ce^{4+} to Ce^{3+} ions.

The performance of soot combustion with Ag$_2$O and Ag doped CeO$_2$ with different silver concentrations was compared at operating temperatures 300–500 °C. The annealing test showed that higher concentration of Ag$^+$ in CeO$_2$:Ag 10:1 catalytic films was as effective as Ag$_2$O films; effectively promoting 100% soot oxidation at T_f = 390 °C due to the oxygen species formed on Ag$^+$ sites. It is worth mentioning that the higher concentrations of Ce^{3+} in the CeO$_2$:Ag 10:1 catalyst films can also form more active oxygen, which then reacts with soot to yield carbon dioxide.

In repetitive tests in loose contact mode, the Ag doped CeO$_2$ catalysts showed stable performance after a small initial decrease (down to approximately 90% of the initial performance for the case of CeO$_2$:Ag 10:1 films) whereas the performance of the pure Ag$_2$O films decreased by 50% after the first test and continued to decrease thereafter.

Overall, the results show that Ag-doped CeO$_2$ films grown by ALD are effective and stable as catalysts for soot oxidation.

Author Contributions: Conceptualization, T.V.I. and A.B., D.C.C.; Methodology, T.V.I., T.H., D.C.C.; Investigation, T.V.I., T.H., A.B., D.C.C.; Writing—Original Draft Preparation, T.V. Ivanova, D.C.C; Supervision, D.C.C.

Funding: This research was funded by TEKES, the Finnish Agency for Technology and Innovation (70006/13), European Regional Development Fund (project CZ.1.05/2.1.00/03.0086), and Ministry of Education Youth and Sports of Czech Republic (project LO1411 (NPU I)).

Acknowledgments: Acknowledgment is given to Monika Stupavská and Pavel Souček for acquiring XPS and XRD data.

Conflicts of Interest: The authors declare no conflict of interest.

References

1. Neeft, J.P.A.; Makkee, M.; Moulijn, J.A. Diesel particulate emission control. *Fuel Process. Technol.* **1996**, *47*, 1–69. [CrossRef]
2. Nielsen, T.; Jørgensena, H.E.; Larsen, J.C.; Poulsen, M. City air pollution of polycyclic aromatic hydrocarbons and other mutagens: Occurrence, sources and health effects. *Sci. Total Environ.* **1996**, *189–190*, 41–49. [CrossRef]
3. Anda, A.; Illes, B. Impact of simulated airborne soot on maize growth and development. *J. Environ. Prot.* **2012**, *3*, 773–781. [CrossRef]
4. Stratakis, G.A.; Stamatelos, A.M. Thermogravimetric analysis of soot emitted by a modern diesel engine run on catalyst-doped fuel. *Combust. Flame* **2003**, *132*, 157–169. [CrossRef]
5. Jequirim, M.; Tshamber, V.; Brilac, J.F.; Ehrburger, P. Oxidation mechanism of carbon black by NO$_2$: Effect of water vapour. *Fuel* **2005**, *84*, 1949–1956. [CrossRef]

6. Pisarello, M.L.; Milt, V.; Peralta, M.A.; Querini, C.A.; Miró, E.E. Simultaneous removal of soot and nitrogen oxides from diesel engine exhausts. *Catal. Today* **2002**, *75*, 465–470. [CrossRef]
7. Prasad, R.; Bella, V.R. A review on diesel soot emission, its effect and control. *Bull. Chem. React. Eng. Catal.* **2010**, *5*, 69–86. [CrossRef]
8. Trovarelli, A.; de Lietenburg, C.; Boaro, M.; Dolcetti, G. The utilization of ceria in industrial catalysis. *Catal. Today* **1999**, *50*, 353–367. [CrossRef]
9. Machida, M.; Murata, Y.; Kishikawa, K.; Zhang, D.; Ikeue, K. On the reasons for high activity of CeO_2 catalyst for soot oxidation. *Chem. Mater.* **2008**, *20*, 4489–4494. [CrossRef]
10. Oliveira, C.F.; Garcia, F.A.C.; Araujo, D.R.; Macedo, J.L.; Dias, S.C.L.; Dias, J.A. Effects of preparation and structure if cerium-zirconium mixed oxides on diesel soot catalytic combustion. *Appl. Catal. A Gen.* **2012**, *413–141*, 292–300. [CrossRef]
11. Krishna, K.; Bueno-Lopez, A.; Makkee, M.; Moulijin, J.A. Potential rare earth modified CeO_2 catalysts for soot oxidation: I. Characterisation and catalytic activity with O_2. *Appl. Catal. B Environ.* **2007**, *75*, 189–200. [CrossRef]
12. Vinodkumar, T.; Rao, B.G.; Reddy, B.M. Influence of isovalent and aliovalent dopants on the reactivity of cerium oxide for catalytic applications. *Catal. Today* **2015**, *253*, 57–64. [CrossRef]
13. Durgasri, D.N.; Vinodkumar, T.; Lin, F.; Alxneit, I.; Reddy, B.M. Gandolinium doped cerium oxide for soot oxidation: Influence of interfacial metal-support interactions. *Appl. Surf. Sci.* **2014**, *314*, 592–598. [CrossRef]
14. Bueno-Lopez, A.; Krishna, K.; Makkee, M.; Moulijn, J.A. Enhanced soot oxidation by lattice oxygen via La^{3+}-doped CeO_2. *J. Catal.* **2005**, *230*, 237–248. [CrossRef]
15. Atribak, I.; Bueno-Lopez, A.; Garcia-Garcia, A. Combined removal of diesel soot particulates and NO_x over CeO_2-ZrO_2 mixed oxides. *J. Catal.* **2008**, *259*, 123–132. [CrossRef]
16. Scire, S.; Riccobene, S.P.M.; Crisafulli, C. Ceria supported group IB metal catalysts for the combustion of volatile organic compounds and the preferential oxidation of CO. *Appl. Catal. B Environ.* **2010**, *101*, 109–117. [CrossRef]
17. Tarasov, A.L.; Przhevalskaya, L.K.; Shvets, V.A.; Kazanskii, V.B. Influence of the metal-oxide electronic interaction on the reactivity of adsorbed oxygen radicals. Applied catalysts containing cerium oxide and Cu, Ag, and Au. *Kinet. Katal.* **1988**, *29*, 1181–1188.
18. Soria, J.; Martinez-Arias, A.; Conesa, J.C. Effect of oxidized rhodium on oxygen adsorption on cerium oxide. *Vacuum* **1992**, *43*, 437–440. [CrossRef]
19. Rangaswamy, A.; Sudarsanam, P.; Reddy, B.M. Rare earth metal doped CeO_2-based catalytic materials for diesel soot oxidation at lower temperatures. *J. Rare Earths* **2015**, *33*, 1162–1169. [CrossRef]
20. Aneggi, E.; Llorca, J.; de Leitenburg, C.; Dolcetti, G.; Trovarelli, A. Soot combustion over silver-supported catalysts. *Appl. Catal. Environ.* **2009**, *91*, 489–498. [CrossRef]
21. Hanedaa, M.; Towata, A. Catalytic performance of supported Ag nano-particles prepared by liquid phase chemical reduction for soot oxidation. *Catal. Today* **2015**, *242*, 351–356. [CrossRef]
22. Kuwahara, Y.; Fujibayashi, A.; Uehara, H.; Mori, K.; Yamashita, H. Catalytic combustion of diesel soot over Fe and Ag-doped manganese oxides: Role of heteroatoms in the catalytic performances. *Catal. Sci. Technol.* **2018**, *8*, 1905–1914. [CrossRef]
23. Shimizu, K.; Kawachi, H.; Satsuma, A. Study of active sites and mechanism for soot oxidation by silver-loaded ceria catalyst. *Appl. Catal. B Environ.* **2010**, *96*, 169–175. [CrossRef]
24. Preda, G.; Pacchioni, G. Formation of oxygen active species in Ag-modified CeO_2 catalyst for soot oxidation: A DFT study. *Catal. Today* **2011**, *177*, 31–38. [CrossRef]
25. Yamazaki, K.; Kayama, T.; Dong, F.; Shinjoh, H. A mechanistic study on soot oxidation over CeO_2-Ag catalyst with 'rice-ball' morphology. *J. Catal.* **2011**, *282*, 289–298. [CrossRef]
26. Castoldi, L.; Aneggi, E.; Matarrese, R.; Bonzi, R.; Llorca, J.; Trovarelli, A.; Lietti, L. Silver-based catalytic materials for the simultaneous removal of soot and NO_x. *Catal. Today* **2015**, *258*, 405–415. [CrossRef]
27. O'Neill, B.J.; Jackson, D.H.K.; Lee, J.; Canlas, C.; Stair, P.C.; Marshall, C.L.; Elam, J.W.; Kuech, T.F.; Dumesic, J.A.; Huber, G.W. Catalyst design with atomic layer deposition. *ACS Catal.* **2015**, *5*, 1804–1825. [CrossRef]
28. Ivanova, T.V.; Toivonen, J.; Homola, T.; Maydannik, P.S.; Kääriäinen, T.; Sillanpää, M.; Cameron, D.C. Atomic layer deposition of cerium oxide for potential use in diesel soot combustion. *J. Vac. Sci. Technol. A* **2016**, *34*, 031506. [CrossRef]

29. Masango, S.S.; Peng, L.; Marks, L.D.; Van Duyne, R.P.; Stair, P.C. Nucleation and growth of silver nanoparticles by AB and ABC-type atomic layer deposition. *J. Phys. Chem. C* **2014**, *118*, 17655–17661. [CrossRef]
30. Puurunen, R.; Vandervorst, W. Island growth as a growth mode in atomic layer deposition: A phenomenological model. *J. Appl. Phys.* **2004**, *96*, 7686. [CrossRef]
31. Shannon, R.D. Revised effective ionic radii and systematic studies of interatomic distances in halides and chalcogenides. *Acta Crystallogr. A* **1976**, *32*, 751–767. [CrossRef]
32. Cui, X.S. *The Basic Theory of Solid Chemistry*; Beijing Institute of Technology Printing: Beijing, China, 1991. (In Chinese)
33. Murray, B.J.; Li, Q.; Newberg, J.T.; Menke, E.J.; Hemminger, J.C.; Penner, R.M. Shape- and size-selective electrochemical synthesis of dispersed silver(I) oxide colloids. *Nano Lett.* **2005**, *5*, 2319–2324. [CrossRef] [PubMed]
34. Tjeng, L.H.; Meinders, M.B.J.; van Elp, J.; Ghijsen, J.; Sawatzky, G.A.; Johnson, R.L. Electronic structure of Ag_2O. *Phys. Rev. B* **1990**, *41*, 3190–3194. [CrossRef]
35. Abe, Y.; Hasegawa, T.; Kawamura, M.; Sasaki, K. Characterization of Ag oxide thin films prepared by reactive RF sputtering. *Vacuum* **2004**, *76*, 1–6. [CrossRef]
36. Kaspar, T.C.; Droubay, T.; Chembers, S.A.; Bagus, P.S. Spectroscopic evidence for Ag(III) in highly oxidized silver films by X-ray photoelectron spectroscopy. *J. Phys. Chem. C* **2010**, *114*, 21562–21571. [CrossRef]
37. Ferraria, A.M.; Carapeto, A.P.; do Rego, A.M.B. X-ray photoelectron spectroscopy: Silver salts revisited. *Vacuum* **2012**, *86*, 1988–1991. [CrossRef]
38. Sadanandam, G.; Kumari, V.D.; Scurrell, M.S. Highly stabilized Ag_2O-loaded nano TiO_2 for hydrogen production from glycerol: Water mixtures under solar light irradiation. *Int. J. Hydrogen Energy* **2017**, *42*, 807–820. [CrossRef]
39. Zhang, H.; Wang, G.; Chen, D.; Lv, X.; Li, J. Tuning photoelectrochemical performances of Ag-TiO_2 nanocomposites via reduction/oxidation of Ag. *Chem. Mater.* **2008**, *20*, 6543–6549. [CrossRef]
40. Xin, B.; Jing, L.; Ren, Z.; Wang, B.; Fu, H. Effects of simultaneously doped and deposited Ag on the photocatalytic activity and surface states of TiO_2. *J. Phys. Chem. B* **2005**, *109*, 2805–2809. [CrossRef] [PubMed]
41. Priya, K.V.B.R.; Shukla, S.; Biju, S.; Reddy, M.L.P.; Patil, K.; Warrier, K.G.K. Comparing ultraviolet and chemical reduction techniques for enhancing photocatalytic activity of silver oxide/silver deposited nanocrystalline anatase titania. *J. Phys. Chem. C* **2009**, *113*, 6243–6255. [CrossRef]
42. Li, J.; Xu, J.; Dai, W.-L.; Fan, K. Dependence of Ag deposition methods on the photocatalytic activity and surface state of TiO_2 with twistlike helix structure. *J. Phys. Chem. C* **2009**, *113*, 8343–8349. [CrossRef]
43. Ansari, A.A.; Labis, J.P.; Alam, M.; Ramay, S.M.; Ahmed, N.; Mahmood, A. Preparation and spectroscopic, microscopic, thermogravimetric, and electrochemical characterization of silver-doped cerium(IV) oxide nanoparticles. *Anal. Lett.* **2017**, *50*, 1360–1371. [CrossRef]
44. Deshpande, S.S.; Patil, S.V.N.; Kuchibhatla, T.; Seal, S. Size dependency variation in lattice parameter and valency states in nanocrystalline cerium oxide. *Appl. Phys. Lett.* **2005**, *87*, 133113. [CrossRef]
45. Cai, S.; Zhang, D.; Zhang, L.; Huang, L.; Li, H.; Gao, R.; Shi, L.; Zhang, J. Comparative study of 3D ordered macroporous $Ce_{0.75}Zr_{0.2}M_{0.05}O_{2-\delta}$ (M = Fe, Cu, Mn, Co) for selective catalytic reduction of NO with NH_3. *Catal. Sci. Technol.* **2014**, *4*, 93–101. [CrossRef]
46. Zou, G.; Fan, Z.; Yao, X.; Zhang, Y.; Zhang, Z.; Chen, M.; Shangguan, W. Catalytic performance of Ag/Co-Ce composite oxides during soot combustion in O_2 and NO_x: Insights into the effects of silver. *Chin. J. Catal.* **2017**, *38*, 564–572. [CrossRef]
47. Bukhtiyarov, V.I.; Havecker, M.; Kaichev, V.V.; Knop-Gericke, A.; Mayer, R.W.; Schlögl, R. Atomic oxygen species on silver: Photoelectron spectroscopy and X-ray absorption studies. *Phys. Rev. B* **2003**, *67*, 235422. [CrossRef]
48. Aneggi, E.; de Leitenburg, C.; Llorca, J.; Trovarelli, A. Higher activity of diesel soot oxidation over polycrystalline ceria and ceria–zirconia solid solutions from more reactive surface planes. *Catal. Today* **2012**, *197*, 119–126. [CrossRef]

© 2018 by the authors. Licensee MDPI, Basel, Switzerland. This article is an open access article distributed under the terms and conditions of the Creative Commons Attribution (CC BY) license (http://creativecommons.org/licenses/by/4.0/).

Article

Electrostatic Supercapacitors by Atomic Layer Deposition on Nanoporous Anodic Alumina Templates for Environmentally Sustainable Energy Storage

Luis Javier Fernández-Menéndez [1,*], Ana Silvia González [1], Víctor Vega [1,2] and Víctor Manuel de la Prida [1,*]

1. Departamento de Física, Facultad de Ciencias, Universidad de Oviedo, C/Federico García Lorca n° 18, 33007 Oviedo, Asturias, Spain; gonzalezgana@uniovi.es (A.S.G.); vegavictor@uniovi.es (V.V.)
2. Laboratorio de Membranas Nanoporosas, Edificio de Servicios Científico Técnicos "Severo Ochoa", Universidad de Oviedo, C/Fernando Bonguera s/n, 33006 Oviedo, Asturias, Spain
* Correspondence: fernandezmluis@uniovi.es (L.J.F.-M.); vmpp@uniovi.es (V.M.d.l.P.); Tel.: +34-985-103-294 (V.M.d.l.P.)

Received: 24 September 2018; Accepted: 10 November 2018; Published: 14 November 2018

Abstract: In this work, the entire manufacturing process of electrostatic supercapacitors using the atomic layer deposition (ALD) technique combined with the employment of nanoporous anodic alumina templates as starting substrates is reported. The structure of a usual electrostatic capacitor, which comprises a top conductor electrode/the insulating dielectric layer/and bottom conductor electrode (C/D/C), has been reduced to nanoscale size by depositing layer by layer the required materials over patterned nanoporous anodic alumina membranes (NAAMs) by employing the ALD technique. A thin layer of aluminum-doped zinc oxide, with 3 nm in thickness, is used as both the top and bottom electrodes' material. Two dielectric materials were tested; on the one hand, a triple-layer made by a successive combination of 3 nm each layers of silicon dioxide/titanium dioxide/silicon dioxide and on the other hand, a simple layer of alumina, both with 9 nm in total thickness. The electrical properties of these capacitors are studied, such as the impedance and capacitance dependences on the AC frequency regime (up to 10 MHz) or capacitance (180 nF/cm^2) on the DC regime. High breakdown voltage values of 60 V along with low leakage currents (0.4 µA/cm^2) are also measured from DC charge/discharge RC circuits to determine the main features of the capacitors behavior integrated in a real circuit.

Keywords: electrostatic supercapacitors; ALD; anodization; nanoporous alumina; energy storage; environmentally sustainable

1. Introduction

Society is increasingly aware of the usage of sustainable energy sources that protect the environment, such as renewable energy resources, whose viability is demonstrated. At the same time, it is necessary that energy storage devices also meet the same requirements of sustainability and efficiency, making the entire process of generation and consumption of energy as clean as possible. Nanotechnology has become one of the main subjects in science nowadays, offering exciting new possibilities in the field of renewable energy production/conversion and storage. An energy storage device is characterized by two main magnitudes. The first one is the energy density, which gives an account of the total energy that the device could store. The other magnitude refers to the time necessary for the device to store or supply a certain amount of energy to attend to the demand, which is known as power density.

Within storage systems, capacitors are those that have higher power densities, but the lowest energy density. Supercapacitors, apart from maintaining the power density of the usual capacitors, are able to reach higher energy density values of about one or two orders of magnitude above those achieved by usual energy storage devices. Therefore, by improving the energy storage capacity of these capacitors, devices with good performance both in autonomy and power density can be obtained, which present a wide range of applications in industry, electronics, or inclusively in means of transport [1,2]. Specifically, supercapacitors have direct application in those electrical systems that demand high power supply in short times, such as the engine of an electric car during vehicle acceleration. Another possible application would be the supply of extra power to the electricity network in specific periods of peak energy demand, which would tackle one of the main disadvantages of renewable energy. In addition, supercapacitors are small gadgets, so it would be possible to introduce this type of device into electronic circuits, causing capacitors to become increasingly smaller and improving their performance due to their high capacitance. Capacitors are usually classified into two types: electrolytic or electrostatic. For the first type, electric charge storage is produced by electrochemical processes. However, the electrolytic acid medium limits the lifespan of these devices as they end up being depleted or even oxidizing the capacitor's own electrodes, rendering it out of service. In addition, these devices have a great disadvantage regarding their ecological footprint, because at the end of their lifetime, they generate non-usable chemical residues. On the other hand, the storage of electrical charge on electrostatic capacitors occurs by the electric polarization of an insulating or dielectric material, when a voltage is applied between the conductive electrodes of the capacitor. From the environmental point of view, their advantages over electrolytic batteries or capacitors are clear: a longer (or even unlimited) lifetime and the non-generation of chemical waste at the end of its working time.

In recent years, several investigations based on nanotechnology have achieved important improvements in energy storage capacity of electrostatic capacitors [3,4]. Manufacturing of devices with high storage capacities and energy densities of up to 4 Wh/kg has been demonstrated [5,6], while maintaining the high power density characteristic of electrostatic capacitors. This type of device is called a super-electrostatic nanostructured capacitor (super-ENC) [7]. The use of nanoporous anodic alumina membranes (NAAMs) as a patterned support for the manufacture of super-ENCs that is proposed in this work is a widely contrasted technique that offers good results not only for the manufacturing of electrostatic capacitors [8], but also for other types of energy storage devices [9,10].

$$C = \varepsilon_0 \varepsilon_r S / d \tag{1}$$

Equation (1) summarizes the main factors to be taken into account when evaluating the storage capacity of an electrostatic capacitor, where C is the capacitance, ε_0 the dielectric permittivity of the vacuum, ε_r the relative permittivity of the dielectric medium of the capacitor, and d is the thickness of such medium. Despite the apparent simplicity of this expression, it can be considered as an approximation to estimate the order of magnitude for a super-ENC's capacitance. As shown in Equation (1), the capacitance is enhanced by a larger surface area (S) of the electrodes. In order to increase the parameter by sizing up the dimensions of the device, NAAMs are used as a substrate, as they offer a high open surface area. By using electrochemical anodization techniques, it is possible to grow hexagonally self-ordered pores whose geometrical parameters can be properly tuned, thus obtaining a highly ordered nanoporous surface with well-defined dimensions, grown on the starting aluminum substrate. The atomic layer deposition (ALD) technique represents an important tool within the nanomaterials additive manufacturing owing to its enhanced possibilities like the fabrication of thin oxide films over micro- or mesoporous surfaces by covering them conformally [11]. The use of the ALD technique allows one to take advantage of the internal surface inside the NAAM's pores to deposit the three layers that make up the typical structure of the capacitor, that is, the upper and the lower electrode and, between them, the dielectric material. The goal of ALD employment is to reduce the thickness of the dielectric material (d) to the order of nanometers, which results in a consequent increase of capacitance (see Equation (1)). This technique allows one to make coatings of oxide materials over porous substrates, achieving the deposition of layers with thicknesses in the range of nanometers on the internal surface of the nanopores [11,12]. Through ALD performance, it is possible to guarantee a uniform thickness substrate coating [13,14], able to be carried out on porous surfaces with diverse morphologies, whether they are micro-, meso-, or macroporous substrates [15]. Within the storage of energy, ALD is a widespread technique and has a wide range of applications depending on the substrate on which the ALD is made, from the treatment of porous carbon to form a cathodic material that can be implemented with batteries [16], to the coating of ZnO nanowires as electrostatic capacitors [17]. This technique is also used to form supercapacitor devices over multiple substrates. There is a detailed study that applies ALD on NAAMs [4–7] in a very similar way to what was conducted in this work, developing new electrostatic supercapacitor devices, in which ALD becomes essential in the conformation of its electrodes and dielectric material. However, its range of application goes further, as ALD has been successfully used in recent years for electrochemical capacitor manufacturing over bundles of carbon nanotubes (CNT) [18,19], or TiO_2 nanoparticles [20]. One of the main advantages provided by ALD is the performance of electrodes with greater mechanical and chemical stability, owing to the deposition of conformal coatings, which are adapted, in many cases, in the form of nanowires [21,22].

In this work, the manufacturing process of electrostatic supercapacitors by combining the ALD technique and electrochemical anodization of nanoporous anodic alumina templates as starting substrates is reported. The so-formed electrostatic capacitor structure, which consists of a top electrode, a dielectric material, and finally a bottom electrode, has been reduced to nanoscale dimensions by depositing the required materials over patterned nanoporous anodic alumina membranes using the ALD technique. A thin layer of aluminum-doped zinc oxide, 6 nm in thickness, is used as both the top and bottom electrodes' material, while two different dielectric materials were tested. On the one hand, a triple-layer made by successive combination of a 3 nm in thickness for each layer of silicon dioxide (SiO_2), plus titanium dioxide (TiO_2), and silicon dioxide (SiO_2) again, forming the three-layered $SiO_2/TiO_2/SiO_2$ dielectric medium, and on the other hand, a single layer with 9 nm in thickness made of alumina (Al_2O_3).

Regarding the selection of dielectric materials, two criteria have been taken into account. Firstly, the material needs to have a high relative permittivity in order to increase the capacitance. The second criterion is related to the energy gap between its valence and conduction bands, which should be as high as possible, to avoid leakage currents and to reach the maximum working voltage of the capacitor. Among the different insulating materials characterized in the work of [23], the one with the highest relative permittivity (TiO_2) and the one with the largest gap (SiO_2) are extracted. Unfortunately, these two properties are opposed, with TiO_2 being the material with the lowest gap, and SiO_2 being the dielectric with the lowest permittivity, as can be seen in Table 1. For this reason, in this work, the combination of these two materials in a triple layer of $SiO_2/TiO_2/SiO_2$ dielectric is proposed, in such a way that this new material can provide the high performance of permittivity and insulation that TiO_2 and SiO_2 display by themselves. The electrical behavior of this new multilayered capacitor is compared with the performance of a single layer capacitor, which is conformed by Al_2O_3 as the dielectric layer. This last dielectric is a material whose ALD deposition has been widely characterized [4–6,11,14,24,25], and which also exhibits an intermediate permittivity and band gap values with respect to TiO_2 and SiO_2.

Table 1. Values of the relative permittivity, ε_r, and band gap of the different materials employed as dielectric layers for the supercapacitors. The given values are those collected from the work of [23].

Dielectric Material	ε_r	Band-Gap (eV)
SiO_2	3.9	9
Al_2O_3	9	8.8
TiO_2	80	3.5

Both oxide materials (Al_2O_3 and $SiO_2/TiO_2/SiO_2$) have 9 nm in total thickness and play the role of insulating layers of the capacitors. The electrical properties of these capacitors were tested under different experimental configurations, by measuring characteristic magnitudes such as the impedance and capacitance on the AC and DC frequency regime. In order to give a complete overview on the electrical behavior of these capacitors, breakdown voltage values along with leakage currents are also studied. This demonstrated the ability of using nanostructured materials for designing energy storage supercapacitor devices.

2. Experimental

2.1. NAAMs Manufacturing

NAAMs have been grown on high-purity (99.999%) Al square plates with 2.5 cm sides and 0.5 mm thickness. The pre-anodization treatment consists of immersing the Al plate in an isopropanol bath and then an ethanol bath, both processes applying ultrasound for 5 min. At this point, a series of electrochemical anodization processes begin, using a Keithley 2400 Source-Meter power supply (Tektronix, Inc., Beaverton, OR, USA). Once washed, electropolishing is carried out in an electrochemical cell for 5 min, applying a DC voltage of 20 V and placing the Al plate as the anodic electrode and a Pt mesh as the cathode. A mixture of perchloric acid and ethanol (1:3 vol) at 5 °C is used as electrolyte, which was mechanically stirred during the process.

The growth of the highly ordered porous alumina is carried out through a double process of electrochemical anodization, as reported in the literature [26,27]. The first anodization is carried out in the electrochemical cell by applying a potentiostatic voltage of 40 V between the electrodes, again placing the aluminum plate on the anodic electrode, and a Pt mesh as the cathode, while mechanically stirring throughout. This process takes 24 h and is performed with a 0.3 M oxalic acid electrolyte at 2 °C. Although during this step, a nanoporous alumina membrane has been produced over the aluminum substrate, the pores of which start to become randomly disordered at the sample surface, it is possible to generate the set of honeycomb self-ordered concavities in the Al substrate. In order to access to the Al substrate, the alumina membrane is then removed through selective chemical etching by an aqueous mixture of chromium trioxide and phosphoric acid for 24 h at room temperature. The second anodization is carried out under the same conditions as the first one, but for a time of 45 min, in which the pores already grow in a hexagonally self-orderly manner, by replicating the honeycomb pattern induced in the substrate by first anodization. Two geometrical parameters of the membrane are fixed when applying this procedure, obtaining pores with a height (h) of 1.2 ± 0.1 µm and placed at a horizontal distance between their centers, or interdistance axes (R_{CC}), of 105 ± 5 nm [4] (Figure 1).

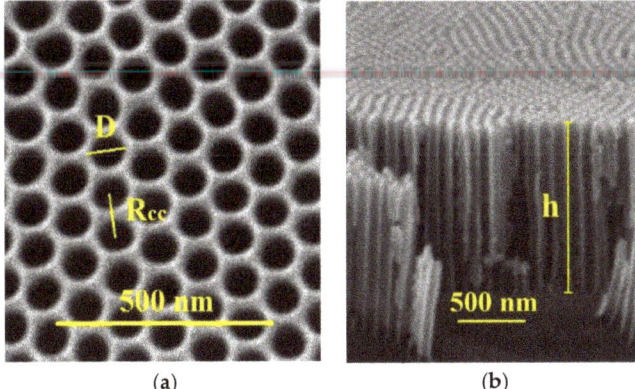

Figure 1. (a) Top view of the nanoporous anodic alumina membrane (NAAM) patterned platform supporting the electrostatic nanostructured capacitor (ENC) structure, showing the honeycomb symmetry acquired by the pores. Geometrical parameters are also shown, such as the distance between centers (R_{CC} = 105 ± 5 nm) and the diameter of the pores (D = 65 ± 3 nm). (b) Side view of the NAAM showing the height (h = 1.2 ± 0.1 µm) of the pores as well as their parallel alignment.

After the second anodization, the diameter of the pores is approximately 35 nm, whereby a widening of the pore size is performed by chemical etching after immersing the membrane in a 5% weight phosphoric acid solution at 30 °C for 35 min. The final average pore diameter (D) of 65 ± 3 nm has been obtained [4]. Thus, the pore radius (r_p) is around 32.5 nm. Figure 1 displays two scanning electron microscope (SEM, JEOL JSM-5600, JEOL Ltd., Akishima, Tokyo, Japan) images of the NAAMs obtained, showing the morphology and geometrical lattice parameters of the patterned alumina membranes employed in this work.

2.2. ALD Performance

One of the foremost features of thermal ALD is that it is mainly limited to the deposition of oxides, so the deposition of a uniform metallic thin film electrode cannot be easily performed [28]. However, it is possible to achieve a semiconductor oxide material layer performance, capable of constituting the electrode material. Then the use of (1:20) aluminum-doped zinc oxide (AZO) is proposed, because its effectiveness as a semiconductor has already been demonstrated for applications in super-ENCs [29], and its deposition performance by ALD has been widely studied [4,29–31].

The ALD process is carried out in a Savannah 100 thermal ALD reactor equipment from Cambridge Nanotech (Waltham, MA, USA), on exposure mode [11], using an Ar flow of 50 sccm as carrier and purge gas. As previously reported [11], a minimum exposure time of 20 s ensures that the precursor gas diffuses properly throughout the substrate, so that the material can be deposited evenly along the entire length of the pores. The number of ALD cycles for each material has been calculated according to the deposition rates shown in Table 2. For example, in the case of the Al_2O_3 dielectric layer, 80 ALD cycles were performed in order to obtain a layer thickness of around 9 nm. The successive layers of material are deposited on the NAAMs sequentially, beginning with the bottom electrode (BE), then the dielectric material, and ending with the top electrode (TE). The method composed by 20 cycles of diethyl zinc (DEZ) intercalated with 1 cycle of trimethyl aluminum (TMA) [4], is used to achieve ZnO doped with Al atoms. This pulse sequence results in an Al doping level of around 3%, which is the optimum that minimizes the resistivity of AZO layers [29]. Thus, the resulting material (AZO) has semiconductor properties and is used as the electrode material in the conductor/dielectric/conductor (CDC) structure of the capacitors, both in the BE and in the TE with a thickness of 6 nm (see Table 2). As already mentioned in the introduction, capacitors have been manufactured employing two different dielectric materials. On the one hand, a single layer of alumina with total thickness of 9 nm, and on the other hand, the multilayered combination of $SiO_2/TiO_2/SiO_2$, in which each layer provides a thickness of 3 nm (see Table 2), again making a total thickness of 9 nm. During the deposition of each material, at least two kind of precursors have been used, with the first of them corresponding to the compound containing the metal, and the other H_2O, which is responsible of the substrate functionalization. For the deposition of SiO_2, it is also necessary to use an O_3 precursor, in order to improve the functionalization performed by H_2O.

Table 2. The different deposited materials are listed, indicating the average deposition rate of each material deposited per atomic layer deposition (ALD) cycle, the chamber temperature during the cycles, the precursor used, as well as the estimated layer thickness for each material. The deposition rates for aluminum-doped zinc oxide (AZO) and alumina have been extracted from the work of [4], while those values for TiO_2 and SiO_2 are obtained from the works of [32,33], respectively. BE—bottom electrode; TE—top electrode.

Material	Deposition Rate	ALD Temperature Reaction (°C)	Precursor	Thickness
AZO	0.19 nm/cycle	200	Diethyl Zinc (DEZ), TMA + H_2O	6 nm (BE/TE)
Al_2O_3	0.13 nm/cycle	200	Trimethyl Aluminum (TMA) + H_2O	9 nm
TiO_2	0.07 nm/cycle	250	Titanium-tetraisopropoxide (TTIP) + H_2O	3 nm
SiO_2	0.06 nm/cycle	180	(3-Aminopropyl) trimethoxysilane (APTES) + H_2O + O_3	3 nm

Each precursor used has its own pulse (t_1), exposure (t_2), and purge (t_3) time, as shown in Table 3. Long exposure (t_2) and purged (t_3) times have been employed, in order to assure that the gaseous precursors have enough time to diffuse into the deep pores.

Table 3. Timing for ALD processes, with t_1 being the precursor pulse time, t_2 the exposition time, and t_3 the purge lapse. For each material, the exposure times used for each precursor are shown by columns. Note that the precursors and times for Al_2O_3 are similar, either used for the conformation of AZO or for the Al_2O_3 dielectric material itself.

Time Period	ZnO		Al_2O_3		TiO_2		SiO_2		
	H_2O	DEZ	H_2O	TMA	H_2O	TTIP	H_2O	O_3	APTES
t_1 (s)	0.1	0.05	0.1	0.05	1	1	1	0.1	2
t_2 (s)	90	90	90	90	60	60	60	60	60
t_3 (s)	180	180	180	180	120	60	120	120	120

In order to ensure the successful deposition of the different layers for the C/D/C capacitor structure, SEM images of the membrane surface have been taken after every deposition step. This characterization allows estimating the thickness of the deposited material attending to the reduction in pores diameter after the placement of each one of the layers that form the capacitor, as can be seen in Figure 2a,c,e. From these images, the homogeneity of the deposited layers can also be appreciated. SEM images have been combined with the EDX technique to study the homogeneity of the materials deposited along the entire pore shape, specifically, for the triple dielectric layer capacitor, as shown in Figure 2b,d,f.

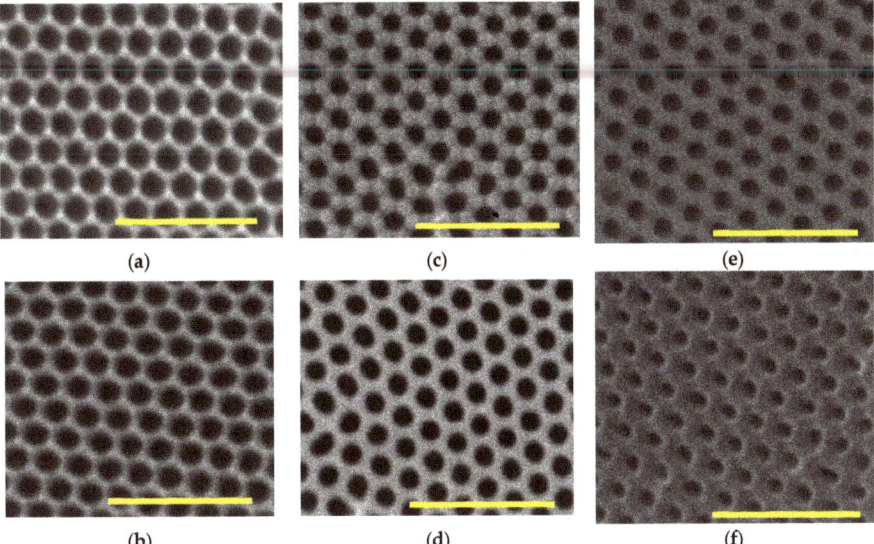

Figure 2. Visualization by scanning electron microscope (SEM) image of the reduction in pores diameter as the layers that make up the capacitor are successively deposited, overlapping one above the other. The left column represents the different stages of the conformation for the single layer capacitor, being (**a**) the micrograph made after the deposition of the BE; (**c**) the one taken after the placement of the Al_2O_3 over the BE; and (**e**) the one that shows the membrane after the deposition of the BE, the Al_2O_3, and the TE. Likewise, the column on the right shows the three similar steps for the triple layer capacitor, being (**b**) the one corresponding to the BE deposition, (**d**) the one corresponding to the triple layer $SiO_2/TiO_2/SiO_2$ deposited over the BE, and (**f**) the one showing the whole BE-$SiO_2/TiO_2/SiO_2$-TE capacitor structure completely deposited. Note the yellow scale bar of 500 nm for all images.

As can be seen in Figure 3, the depth profiles for Al (blue trace), Si (red), and Ti (green) remain stable, indicating that a uniform coating along the whole pores size has been carried out.

Figure 3. SEM image of the cross section for the NAAM with the deposited triple dielectric layer capacitor. On the left side, the substrate of Al can be seen, while on the right side, the NAAM surface appears where the pores are opened. EDX analysis has been carried out along the yellow segment, indicating the presence of different elements through it. The blue line corresponds to Al, while the red and green ones correspond to Si and Ti, respectively.

Using ImageJ software (version 1.52a, National Institutes of Health, Bethesda, MD, USA), the reduction in pore diameter has been calculated from the SEM surface images, and the results are shown in Table 4. Taking into account the starting diameter of 65 nm and the thickness values shown in Table 2, the pore diameter is expected to be around of 53 nm after the deposition of the BE, 35 nm after the dielectric material conformation, and 23 nm after deposition of the TE. The experimental data obtained for the reduction in pore diameter are in good agreement with expectations, so that the respective layers forming the capacitor have the appropriate thicknesses.

Table 4. Average pore diameters after deposition of the different layers, either BE, dielectric, or TE. Data have been obtained from the analysis of scanning electron microscope (SEM) images with ImageJ software.

Deposition Stage	Pore Diameter of Single Dielectric Layer Capacitor (nm)	Pore Diameter of Triple Dielectric Layer Capacitor (nm)
BE	52 ± 3	51 ± 5
BE + Dielectric	33 ± 4	38 ± 3
BE + Dielectric + TE	27 ± 3	20 ± 2

2.3. Electrical Contacts

The electrical contacts were made by fixing a copper wire with silver paint directly on the two electrodes of the capacitor. To isolate the electrodes and thus be able to access the bottom one without limitations, part of the NAAM has been masked with a Kapton tape after the BE deposition process. Once the deposition of the dielectric material and the TE is done, the Kapton mask is removed with acetone in such a way that the lower electrode can be further contacted.

2.4. Electrical Characterization

To describe the properties of the capacitors, the electrical behavior of these devices has been analyzed by integrating them on a real electronic circuit. In order to do this, two stages have been differentiated according to the nature of the current supplied to the device. First, the properties have been studied in a dynamic regime, that is, by supplying an alternating current, from 40 Hz to 100 MHz on frequency range. In this work, a precision impedance analyzer model Agilent 4294A (Keysight Technologies, Santa Rosa, CA, USA) has been used. This study allows knowing the characteristic magnitudes of the capacitor, such as the impedance module and phase as a function of the input AC frequency. Secondly, the capacitors were tested in static regime, that is, with continuous input current. This section is essential when checking the energy storage properties of the device because the procedure consists of charging and discharging the device, analyzing the intensity of the current flowing through the capacitor at all times. To allow this, the assembly of a charge–discharge RC circuit is required, as shown in Figure 4.

V represents the DC voltage supplied by the power supply, R_L is the load resistance, R_D is the discharge resistance, and the values are set as $V = 2$ V and $R_L = R_D = 2$ MΩ. In addition, the circuit is fitted with an ammeter (A) to measure the current intensity (I) flowing through the capacitor branch as well as a switch (S) to select the working mode of the capacitor (C), either charging or discharging. A Keithley 2410 1100 V Source-Meter (Tektronix, Inc., Beaverton, OR, USA) has been used as the power source and a Keithley 2700 1100 V Multimeter/Data Acquisition System (Tektronix, Inc., Beaverton, OR, USA) as the ammeter. Through the data of current intensity transient collected by this device, it is possible to know the energy storage capacity of the capacitor.

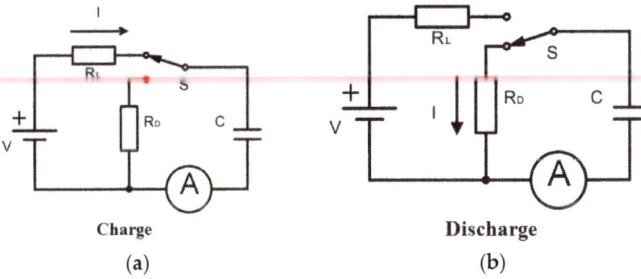

Figure 4. Schematic view of the RC charge–discharge circuit. In the charge configuration, (**a**) the capacitor (C) is powered by the power supply (V) through the load resistance ($R_L = 2$ MΩ). In the discharge configuration, (**b**) the discharge resistance (R_D) is powered by the capacitor. Note that the current (I) flows in different directions through the ammeter (A) depending on whether the configuration is charge or discharge.

By modifying the circuit (see Figure 5), it is possible to find the maximum operating voltage of the device or breakdown voltage, in which the dielectric medium loses its insulating properties and the current circulates through it as if it were a conductive medium, thus losing the properties for energy storage.

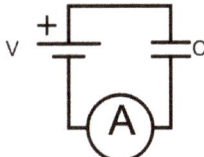

Figure 5. Circuit diagram for breakdown voltage measurements. The power supply (V) is directly connected to the capacitor (C), so the current flowing through the device can be simply monitored by the ammeter (A). A linear increase in the current read by the ammeter indicates the loss of the insulating properties of the capacitor.

3. Mathematical Framework

3.1. Capacitance Estimation

In parallel with the development of the super-ENCs manufacturing procedures, analytical calculations have also been carried out to estimate the electrical properties that can be expected from these devices. Special attention has been paid to magnitudes as the capacitance, the behavior of the phase, and the impedance module when an AC current is applied, as well as the current flowing through the device when in use.

To estimate the capacitance of the device, it is essential to know the symmetry of the substrate on which the layers that make up the super-ENC are deposited. It is considered that inside each one of the NAAM's pores, a capacitor of cylindrical symmetry is generated, all of them with similar characteristics because of the geometrical uniformity of the NAAM. This single capacitor is connected in parallel with the six single capacitors of cylindrical symmetry present in the six adjacent pores, as a result of the spatial pores distribution with hexagonal symmetry. The total capacitance of the device is then the sum of the capacitance of all the individual capacitors present in the NAAM, as they are connected in parallel with each other. In Figure 6, a diagram of the cross-section of the capacitor generated within a pore can be seen. Three parts can be clearly differentiated according to the geometry, each of them associated with a type of capacitor. In the upper part, a flat symmetry capacitor appears; the intermediate part corresponds to a capacitor of cylindrical symmetry; and the lower part, the bottom of the pore, is associated with a capacitor of hemispherical symmetry. From the calculations reported in the work of [5], where the capacitance provided by each of the parts is calculated, it follows that the main contribution to the total capacitance of the super-ENC comes from the cylindrical part of the pores. Taking into account that the pores of the NAAMs used in this work have a height of 1.2 µm, it is estimated that 95% of the capacitance of the super-ENC comes from the cylindrical part, so the contributions of the top flattened and bottom hemispherical parts are considered negligible.

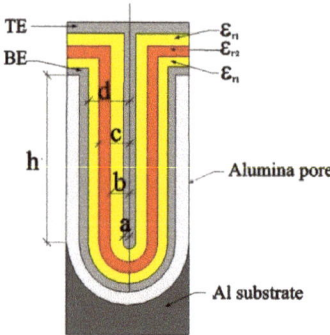

Figure 6. Schematic cross-section of the supercapacitor conductor/dielectric/conductor (C/D/C) structure across a NAAM's pore. All the sections that make up the device are represented, starting from the aluminum substrate on which the pores are grown. Between the electrodes (TE and BE), the layers forming the dielectric material can be seen, representing those corresponding to the triple layer capacitor in this figure. The yellow color represents the SiO_2 layers and the orange represents the one of TiO_2, indicating their relative permittivity (ε_{R1} and ε_{R2}) and thickness (a, b, c, and d). Note that for the single layered capacitor, instead of three dielectric layers, there would be only one, occupying the same space as the sum of the three layers.

One of the innovations presented in this work relays on the estimation of the capacitance for triple dielectric layer capacitors ($SiO_2/TiO_2/SiO_2$) instead of single layer capacitors. This fact constitutes the main contribution of this work with respect to the calculations exposed in previous literature [5], where only single layer capacitors are treated. The capacitance expression for the cylindrical part is then slightly different, as shown in Equation (2), where a, b, c, and d represent the radii of the different dielectric layers with respect to the axis of symmetry of the cylinder. Therefore, the interdistance $d - c$ corresponds to the thickness of the internal layer, $c - b$ to that of the intermediate layer, and $b - a$ to that of the outer layer of dielectric. Note that a is the distance from the axis of the cylinder to the outer layer of dielectric, a quantity that does not depend on the thickness of the TE. On the other hand, $r_p - d$ is the thickness related to the BE, but this one does not influence the capacitance calculations. As the internal and external dielectric layers are formed by the same material, two values of relative permittivity come into play, ε_{r1} for the inner and outer layers (SiO_2) and ε_{r2} for the intermediate layer (TiO_2). In the case of the super-ENC formed by a single layer (Al_2O_3), the capacitance calculation for a single pore is simpler (Equation (3)), obtaining an expression similar to that shown in the work of [5]. For the single layer capacitor, the thickness and relative permittivity of the dielectric layer is given by $d - a$ and ε_r, respectively.

$$C = \frac{2\pi\varepsilon_0 h}{\frac{1}{\varepsilon_{r_1}} \ln\left(\frac{db}{ca}\right) + \frac{1}{\varepsilon_{r_2}} \ln\left(\frac{c}{b}\right)} \tag{2}$$

$$C = \frac{2\pi\varepsilon_0 \varepsilon_r h}{\ln(d/a)} \tag{3}$$

To account for the total capacitance of the super-ENC, the capacitance density is usually calculated, that is, the normalized capacitance per unit area of the NAAM. It is necessary, therefore, to know the number of pores per unit area (σ) that the NAAMs present, which is given by Equation (4), where the hexagonal symmetry of the membranes is also taken into account.

$$\sigma = \frac{2}{\sqrt{3}R_{CC}^2} \tag{4}$$

Then, the capacitance density of supercapacitors can be obtained as the result of multiplying Equations (4) and (2) or (3) (for tri-layered or single-layered dielectric material, respectively). By introducing a, b, c, and d distances according to the thicknesses of the respective dielectric layers (see Table 2), the geometrical parameters of the NAAMs, R_{CC}, and h (see Figure 1 in the Manufacturing section), as well as the relative permittivity values, ε_r, (which are shown in Table 1), it is possible to estimate the capacitance densities for the manufactured super-ENCs (Table 5).

Because manufactured NAAMs have a surface area of 0.7 cm^2, the expected capacitances for super-ENCs would be around of 6.7 and 14.8 µF for the triple and single dielectric layer capacitors, respectively.

Table 5. Distances from the pore axis to the different dielectric layers and estimation of the capacitance density for the manufactured devices.

Dielectric Material	a (nm)	b (nm)	c (nm)	d (nm)	Capacitance Density (F/cm^2)
Triple layer (SiO$_2$/TiO$_2$/SiO$_2$)	17.5	20.5	23.5	26.5	9.5
Single layer (Al$_2$O$_3$)	17.5	–	–	26.5	21.1

3.2. Electrical Behavior of the Capacitor in a Real Circuit

Inserting an electronic device in a real circuit is the most appropriate approach to check its operational behavior. In particular, by monitoring the current passing through the capacitor as a function of time, two magnitudes that characterize the device can be known, such as its capacitance (C) and leakage current (I_{LK}). These two features can be measured at the same time with a given experimental configuration, which is the load curve of the capacitor within a DC RC circuit (see Figure 4a). To extract the appropriate information, Equation (5) has been deduced, which represents the decay of the current (I) that the ammeter records as function of time (t), from a maximum current value (I_0). As can be seen in Figure 7, and from Equation (5) also, a resistance in parallel to the capacitor has been included in such a way that this element represents the internal resistance of the capacitor itself to the current flow, which is, the leakage resistance (R_{LK}). Note that this resistance is not a real element of the experimental circuit.

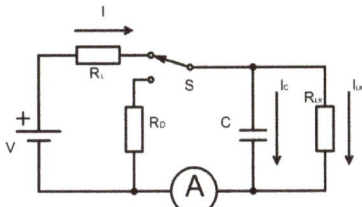

Figure 7. Theoretical RC charge circuit setup. The ideal model that simulates the electrical behavior of the capacitor (C) in the charging mode has been obtained by adding a parallel resistance, or leakage resistance (R_{LK}), in the capacitor's branch. The intensity (I) flowing through the load resistance (R_L) would be divided into two, the one that circulates through the capacitor itself (I_C), and the one circulating through the leak resistance (I_{LK}). Note that this figure is a schema because the real circuit used is the one represented in Figure 4a.

Equation (5) shows the intensity flowing through the ammeter as the sum of an intensity that decays due to the capacitance presence (I_C), plus the capacitor leakage current, which has a constant value and acts as an offset. This magnitude varies according to the DC voltage (V) applied by the source and the load resistance (R_L) of the circuit, as can be seen in Equations (6) and (7). Then, the leakage intensity is a relative property, typical of the circuit or experimental configuration, so this work is also going to account for an intrinsic feature of the capacitor, as it is the leakage resistance. Note also that this model is valid as long as R_L is much greater than the internal resistance (R) of the capacitor itself, otherwise R should be taken into account as a series resistor in the capacitor's branch. The real capacitor could be thus modelled as an RC circuit (where the R value should be the capacitor's internal resistance and C the capacitance) connected to a resistor in parallel, which represents the leakage resistance. As will be demonstrated below, the internal resistance of the capacitor (R) has a value three orders of magnitude lower than the load resistance (R_L), so the presence of the internal capacitor's resistance can be eliminated in the theoretical charge–discharge circuit (as shown in Figure 7).

$$I(t) = \frac{V}{R_{LK} + R_L} + I_0 e^{-\frac{R_{LK}+R_L}{R_{LK}R_L}\frac{1}{C}t} \tag{5}$$

$$I_{LK} = \frac{V}{R_{LK} + R_L} \tag{6}$$

$$I_C(t) = I_0 e^{-\frac{R_{LK}+R_L}{R_{LK}R_L}\frac{1}{C}t} \tag{7}$$

4. Results and Discussion

4.1. Dynamic Regime Study

Next, the analysis performed with the manufactured capacitor working under AC conditions will be explained. The data obtained from the impedance analyzer are the impedance module ($|Z|$) and the capacitor phase as a function of the applied current frequency. As can be seen in Figure 8a,b, both the single layer capacitor as well as the triple layer capacitor show a decay of the impedance module until reaching the resonance frequency, which is located around 2.9×10^7 Hz for the single-layered capacitor, and around 1.7×10^7 Hz for the triple-layered capacitor. Being capacitive electronic components, the output signal of this type of devices has a phase shift of $-90°$ with respect to the input current. However, once the resonance frequency has been exceeded (in which the imaginary part of the impedance is canceled), this value changes to positive $90°$, becoming in an inductive element. This pattern can be clearly seen in Figure 8a,b.

$$|Z| = \sqrt{R^2 + \left(\frac{1}{2\pi v C}\right)^2} \tag{8}$$

Despite that the phase of the manufactured devices does not remain constant at $-90°$ throughout the frequency range analyzed, it remains close to this value within the low frequency range, confirming that manufactured devices behave as capacitors in this region. In fact, this work places special emphasis on the electrical behavior at a low frequency, as its application for electrostatic energy storage requires the use of low frequency input signals. Following this approach, it is considered that, at low frequency, the capacitor can be analyzed as a pure RC circuit. The data of the impedance module as a function of the frequency (v) have been fitted to Equation (8), in such way that R represents the internal resistance of the device and C its capacitance. Note that all the fittings shown below have been made by least squares, offering a minimum value for R^2 of 0.999, while fitting uncertainties have been calculated with a 95% confidence level. In Figure 8c,d, it is possible to see how, between 40 and 1000 Hz, the electrical behavior of the manufactured capacitors are similar to an RC circuit, because fittings exactly represent the experimental data. R and C values have been extracted from the fitting for each type of capacitor, as shown in Table 6.

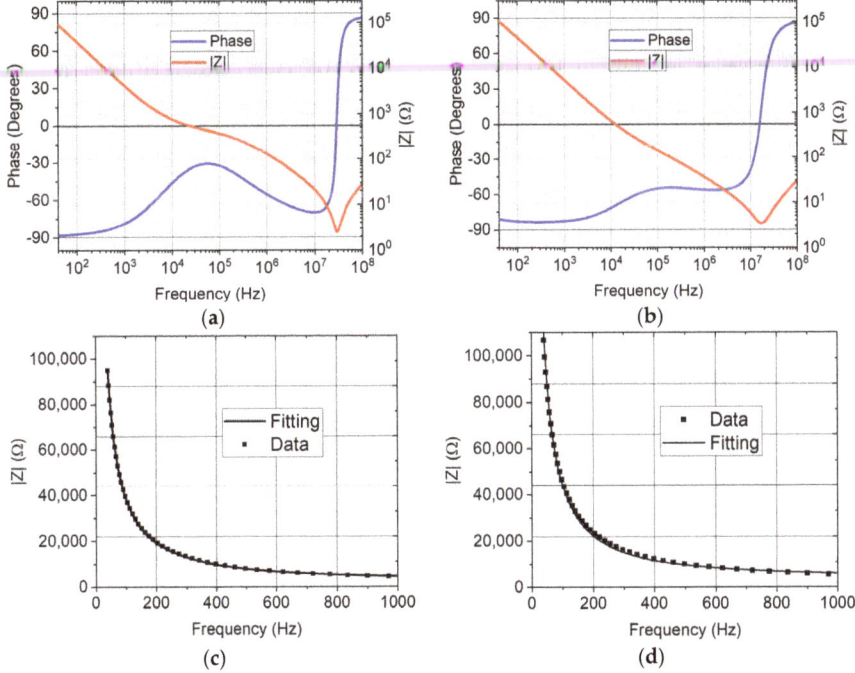

Figure 8. Module of impedance (|Z|) and phase curves as a function of frequency for the single dielectric layer (a) and triple dielectric layer (b) capacitors. Below, the impedance module data limited to the low-frequency range (from 40 up to 1000 Hz) and its fittings for the single dielectric layer (c) and triple dielectric layer capacitor (d), respectively, are represented.

By considering that R is the internal resistance of the capacitor and taking into account that it reaches values between 2.3 and 3.9 kΩ, they are of in the order of 1000 smaller than the 2 MΩ for the load resistance. In this way, the condition imposed in the mathematical framework section is fulfilled in order to apply Equation (5) to the charge curves.

Table 6. Main features obtained from low frequency impedance fitting curves of the manufactured capacitors.

Dielectric Material	Resistance (R)	Capacitance (C)
Single layer (Al$_2$O$_3$)	2.3 ± 0.2 kΩ	41.60 ± 0.07 nF
Triple layer (SiO$_2$/TiO$_2$/SiO$_2$)	3.9 ± 0.5 kΩ	36.2 ± 0.2 nF

4.2. Static Regime Study

After an analysis of the capacitor's properties under an AC input current, another test performed is reported, this time applying an input current DC through the devices. In particular, the charge–discharge cycles have been analyzed on a test circuit (see Figure 4) where the super-ENC prototypes have been placed, obtaining the intensity signal for the capacitor's branch as a function of time. The typical signal of the charge–discharge cycles in circuits including this type of electrostatic capacitors has two main characteristics; namely, the intensity decays are symmetrical with respect to the time axis, and the curves have different signs. The symmetry is due to the fact that R_L and R_D have the same value, and the sign differences are caused by the polarity of the capacitor's branch, which is inverted depending on whether it is in charge mode or in discharge mode. In this work, these conditions are met because R_L and R_D have a value of 2 MΩ, however, as can be seen in Figure 9, the measured signals have a particularity because the charge cycles appear elevated by a constant value with respect to the 0. This offset is of special interest as it accounts for the leakage current of the capacitor (represented by a red line).

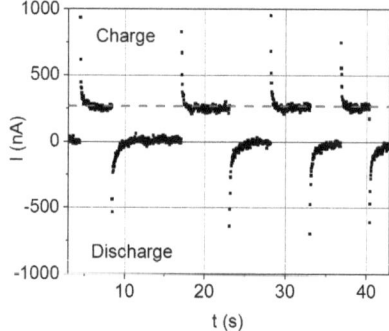

Figure 9. Several charge and discharge cycles for the triple dielectric layer capacitor, representing the intensity (*I*) flowing through the ammeter of the circuit as a function of time (*t*). An offset current (red dashed line) can be observed for charge cycles, revealing the leakage current of the capacitor.

By applying Equation (5) to the load cycle, not only *C* can be estimated, but also both I_{LK} and R_{LK} can be determined. To improve the quality of the fitting as well as the statistics of the experiment, the signal of three consecutive loads is accumulated in order to fit a curve that contains triple the number of points. As can be seen in Figure 10, the proposed model is fitted to the experimental results, which is shown in Table 7. It should be noted that for the single dielectric layer capacitor, it was not possible to detect the presence of leakage current, being, in the existing case, below 50 nA, which is the minimum resolution of the experiment.

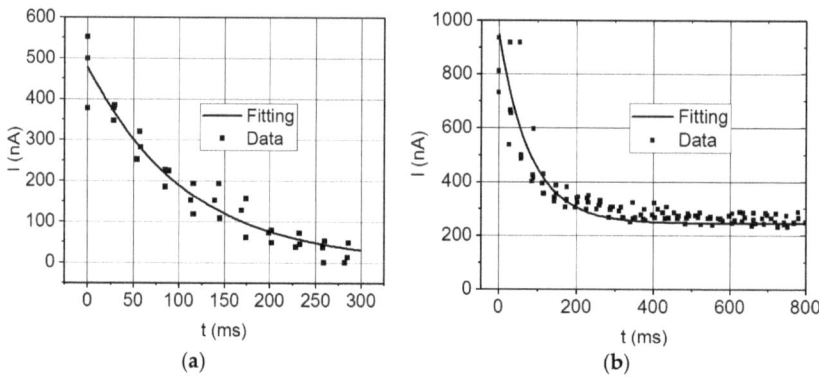

Figure 10. Intensity (*I*) decays in charge cycle for single dielectric layer (**a**) and triple dielectric layer (**b**) capacitors as a function of time (*t*). Its fittings to Equation (5) are also shown.

Table 7. Mean results obtained by applying Equation (5) to the charge cycle intensity decays.

Dielectric Material	Leakage Current (I_{LK})	Leakage Resistance (R_{LK})	Capacitance (C)
Single layer (Al_2O_3)	–	–	54 ± 5 nF
Triple layer ($SiO_2/TiO_2/SiO_2$)	220 ± 60 nA	7.1 ± 0.8 MΩ	93 ± 8 nF

4.3. Breakdown Voltage Test

Finally, the maximum voltage value to which the device can operate was tested. For this, the device is placed in the circuit of Figure 5 and the intensity shown by the ammeter is recorded as a function of the applied voltage. The triple dielectric layer capacitor ($SiO_2/TiO_2/SiO_2$) exhibited a breakdown voltage of 14 ± 1 V, while the single dielectric layer capacitor (Al_2O_3) has reached a higher value at 63 ± 1 V. In Figure 11, the voltage ranges at which the capacitors lose their insulating properties and become conductors appear highlighted, as they are indicated by an arrow. It is clear that the multi-layered dielectric $SiO_2/TiO_2/SiO_2$ is not reaching the insulating features that it was supposed display. On the one hand, it exhibits leakage currents and on the other hand, it has a lower breaking voltage than the Al_2O_3 capacitor. As no inhomogeneities have been detected in the triple dielectric layer of $SiO_2/TiO_2/SiO_2$ ALD deposited material, the main causes of this decrease in performance may be other reasons. For example, a higher ALD process temperature for SiO_2 causes breaking voltage decreasing [34]. It can also be because of the combination of thin films depositions of different oxides, which may form a new alloyed material that leads to a band-gap reduction with respect to the corresponding ones of SiO_2 and TiO_2 [35]. In such a way, an a priori insulating material becomes a semiconductor material and, consequently, it could not perform as a dielectric medium.

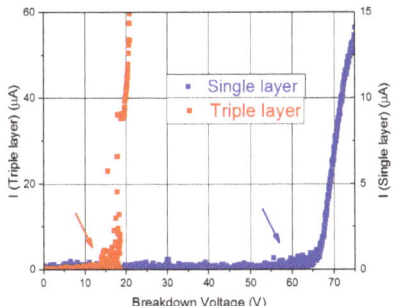

Figure 11. Breakdown current–voltage curves for the single dielectric layer (red) and three-layered (blue) capacitors. Representing the intensity (*I*) recorded by the ammeter (Figure 5) as a function of the applied voltage, it is possible to find the breakdown values, which are indicated by an arrow.

5. Conclusions

This work has faced the development of electrostatic capacitors and its enhanced possibilities by using nanomaterials, in this way covering the full manufacturing and characterization process of these energy storage devices. An innovative fabrication method has been proposed and achieved, based on the successive combination of an ultrathin layered nanomaterial for the conformation of the dielectric medium of the capacitors. Likewise, an experimental procedure has been followed for the complete characterization of these devices, consisting of three phases, from which the intrinsic magnitudes that completely characterize a capacitor can be measured, such as internal resistance, leakage resistance, capacitance, and breakdown voltage. The test of the manufactured devices in a real circuit, including a model to explain their electrical behavior, which is the main novelty of this study, has obtained experimental results confirming the validity of such a model.

It has been found that the Al_2O_3 single-dielectric layer capacitors of 9 nm in thickness have been shown to exhibit a better performance than the triple dielectric layer capacitors composed of $SiO_2/TiO_2/SiO_2$ sheets, each 3 nm in thickness. In particular, single dielectric layer capacitors have less internal resistance (2.3 kΩ), so they are more favorable for storage applications because of a consequent lower power consumption. Al_2O_3 single-layered capacitors also offer a higher capacitance in dynamic regime (41.6 nF) than those of triple-layered $SiO_2/TiO_2/SiO_2$. Furthermore, the former display leakage current is lower than 50 nA, so it guarantees that the current losses will be minimal. The main advantage shown by these capacitors is the high value of breakdown voltage (63 V), as a higher working voltage greatly improves the storage capacity of electrical energy of these devices. Only the static regime of capacitance for triple dielectric layer capacitors (93 nF) is higher than that of the single dielectric layer devices (54 nF). However, taking into account all the features in which the Al_2O_3 capacitors offer better performance, a single property is not enough to affirm that $SiO_2/TiO_2/SiO_2$ devices have better characteristics.

Nevertheless, the capacitance values derived from our electrostatic supercapacitor prototype are not as high as expected. The values obtained are coherent with the performed analysis, as both the dynamic and the static procedures yield values of the same order of magnitude (nF). However, they are far from the expected capacitance values in the µF range theoretically predicted for these kind of devices, so certain aspects of the manufacture of the supercapacitor devices should be further reconsidered. A feasible explanation on the discrepancies between the expected values of capacities and the experimentally measured ones is that the AZO layer is not properly fulfilling its function as electrode material, for either of two reasons. The first one is that the contact to the AZO layer with silver paint would not be appropriate and thus there is no electron transfer between the AZO layer and the conductive silver paint. The second reason would be that the AZO layer itself is a semiconducting material, and hence it is not able to efficiently conduct the electrical current along the channels of the pores. In addition, the design of the electrical connections of the device becomes critical, because current leakages and short circuits between the electrodes need to be avoided. For example, leakage currents may be decreased in the case of triple layer capacitors, whether or not it can be guaranteed that the electrodes are completely isolated to achieve the most desirable device performance.

There are, therefore, two ways of improving the super-ENCs' capacitance. On the one hand, the substitution of the AZO layer by using a better conductive material that fits the cylindrical morphology of the pores, such as carbon nanotubes, thus taking advantage of the internal surface of the NAAM to increase the capacitance of the devices. On the other hand, the use of more refined techniques to contact the electrodes, such as wire bonding, would allow precise delimitation of the contact zones, avoiding regions in which short circuits could occur, thus reducing the presence of leakage currents. These advances would significantly improve the performance of the manufactured prototypes that, according to the reported results, could become in very promising energy storage devices. In fact, the super-ENCs are suitable complements for the batteries of electrical systems such as vehicles or electricity supply domestic networks. Besides having a high energy density and faster response under a specific power demand, they also present an environmentally sustainable alternative to the current polluting energy supply systems.

Author Contributions: Conceptualization, A.S.G. and V.M.d.l.P.; Funding Acquisition: V.M.d.l.P. and V.V.; Investigation, L.J.F.-M., A.S.G., and V.V.; Writing—Original Draft Preparation, L.J.F.-M. and A.S.G.; Writing—Review & Editing, A.S.G., V.M.d.l.P., L.J.F.-M., and V.V.; Supervision, V.M.d.l.P.

Funding: This work was funded by the Ministerio de Economía y Competitividad (Spain) under grant MINECO-17-MAT2016-76824-C3-3-R, and IDEPA-ArcelorMittal Proof of Concept grant No RIS3-2015: SV-PA-15-RIS3-2.

Acknowledgments: Funding by Spanish MINECO under grant No MAT2016-76824-C3-3-R is gratefully recognized. The scientific support from the SCTs of the University of Oviedo is acknowledged. Helpful assistance within the electrical measurements provided by the group of Sistemas Electrónicos de Alimentación (SEA) led by J. Sebastián Zuñiga from the University of Oviedo is also gratefully recognized. This work is devoted to the memory of our colleague and friend J. Miguel Mesquita Teixeira.

Conflicts of Interest: The authors declare no conflict of interest.

References

1. Sakka, M.A.; Gualous, H.; Omar, N.; Mierlo, J.V. Batteries and supercapacitors for electric vehicles. *New Gener. Electr. Veh.* **2012**, *5*, 134–160. [CrossRef]
2. Devillers, N.; Jemei, S.; Péra, M.-C.; Bienaimé, D.; Gustin, F. Review of characterization methods for supercapacitor modelling. *J. Power Source* **2014**, *246*, 596–608. [CrossRef]

3. Wang, Y.; Guo, J.; Wang, T.; Shao, J.; Wang, D.; Yang, Y.-W. Mesoporous transition metal oxides for supercapacitors. *Nanomaterials* **2015**, *5*, 1667–1689. [CrossRef] [PubMed]
4. Iglesias, L.; Vega, V.; Garcia, J.; Hernando, B.; Prida, V.M. Development of electrostatic supercapacitors by atomic layer deposition on nanoporous anodic aluminum oxides for energy harvesting applications. *Front. Phys.* **2015**, *3*, 12. [CrossRef]
5. Banerjee, P.; Perez, I.; Henn-Lecordier, L.; Lee, S.B.; Rubloff, G.W. Nanotubular metal-insulator-metal capacitor arrays for energy storage. *Nat. Nanotechnol.* **2009**, *4*, 292–296. [CrossRef] [PubMed]
6. Haspert, L.C.; Lee, S.B.; Rubloff, G.W. Nanoengineering strategies for metal-insulator-metal electrostatic nanocapacitors. *ACS Nano* **2012**, *6*, 3528–3536. [CrossRef] [PubMed]
7. Pérez, I. Nanoporous AAO: A Platform for Regular Heterogeneus Nanostructures and Energy Storage Devices. Ph.D. Thesis, University of Maryland, College Park, MD, USA, August 2009.
8. Han, F.; Meng, G.; Zhou, F.; Song, L.; Li, X.; Hu, X.; Zhu, X.; Wu, B.; Wei, B. Dielectric capacitors with three-dimensional nanoscale interdigital electrodes for energy storage. *Sci. Adv.* **2015**, *1*, e1500605. [CrossRef] [PubMed]
9. Zhao, H.; Liu, L.; Lei, J. A mini review: Functional nanostructuring with perfectly-ordered anodic aluminum oxide template for energy conversion and storage. *From Chem. Sci. Eng.* **2018**, *12*, 481–493. [CrossRef]
10. Li, J.; Cheng, X.; Shashurin, A.; Keidar, M. Review of electrochemical capacitors based on carbon nanotubes and graphene. *Graphene* **2012**, *1*, 1–13. [CrossRef]
11. Elam, J.W.; Routkevitch, D.; Mardilovich, P.P.; George, S.M. Conformal coating on ultrahigh-aspect-ratio nanopores of anodic alumina by atomic layer deposition. *Chem. Mater.* **2003**, *15*, 3507–3517. [CrossRef]
12. George, S.M. Atomic layer deposition: An overview. *Chem. Rev.* **2010**, *110*, 111–131. [CrossRef] [PubMed]
13. Yuan, G.; Wang, N.; Huang, S.; Liu, J. A brief overview of atomic layer deposition and etching in the semiconductor processing. In Proceedings of the 17th International Conference on Electronic Packaging Technology, ICEPT 2016, Wuhan, China, 16–19 August 2016; pp. 1365–1368.
14. Vega, V.; Gelde, L.; González, A.S.; Prida, V.M.; Hernando, B.; Benavente, J. Diffusive transport through surface functionalized nanoporous alumina membranes by atomic layer deposition of metal oxides. *J. Ind. Eng. Chem.* **2017**, *52*, 66–72. [CrossRef]
15. Daubert, J.S.; Wang, R.; Ovental, J.S.; Barton, H.F.; Rajagopalan, R.; Augustyn, V.; Parsons, G.N. Intrinsic limitations of atomic layer deposition for pseudocapacitive metal oxides in porous electrochemical capacitor electrodes. *J. Mater. Chem. A* **2017**, *5*, 13086–13097. [CrossRef]
16. Lei, Y.; Lu, J.; Luo, X.; Wu, T.; Du, P.; Zhang, X.; Ren, Y.; Wen, J.; Miller, D.J.; Miller, J.T.; Sun, Y.-K.; Elam, J.W.; Amine, K. Synthesis of porous carbon supported palladium nanoparticle catalysts by atomic layer deposition: Application for rechargeable lithium–O_2 battery. *Nano Lett.* **2013**, *13*, 4182–4189. [CrossRef] [PubMed]
17. Wei, L.; Liu, Q.-X.; Zhu, B.; Liu, W.-J.; Ding, S.-J.; Lu, H.-L.; Jiang, A.; Zhang, D.W. Low-cost and high-productivity three-dimensional nanocapacitors based on stand-up ZnO nanowires for energy storage. *Nanoscale Res. Lett.* **2016**, *11*, 213. [CrossRef] [PubMed]
18. Fiorentino, G.; Vollebregt, S.; Tichelaar, F.D.; Ishihara, R.; Sarro, P.M. 3D solid-state supercapacitors obtained by ALD coating of high-density carbon nanotubes bundles. In Proceedings of the IEEE International Conference on Micro Electro Mechanical Systems (MEMS), San Francisco, CA, USA, 26–30 January 2014; pp. 342–345. [CrossRef]
19. Kao, E.; Yang, C.; Warren, R.; Kozinda, A.; Lin, L. ALD titanium nitride on vertically aligned carbon nanotube forests for electrochemical Supercapacitors. *Sens. Actuators A* **2016**, *240*, 160–166. [CrossRef]
20. Hai, Z.; Karbalaei Akbari, M.; Wei, Z.; Xue, C.; Xu, H.; Hu, J.; Hyde, L.; Zhuiykov, S. TiO_2 nanoparticles-functionalized two-dimensional WO_3 for high-performance supercapacitors developed by facile two-step ALD process. *Mater. Today Commun.* **2017**, *12*, 55–62. [CrossRef]
21. Zheng, W.; Cheng, Q.; Wang, D.; Thompson, C.V. High-performance solid-state on-chip supercapacitors based on Si nanowires coated with ruthenium oxide via atomic layer deposition. *J. Power Sour.* **2017**, *341*, 1–10. [CrossRef]
22. Wang, R.; Xia, C.; Wei, N.; Alshareef, H.N. $NiCo_2O_4$@TiN core-shell electrodes through conformal atomic layer deposition for all-solid-state supercapacitors. *Electrochim. Acta* **2016**, *196*, 611–621. [CrossRef]
23. Robertson, J. High dielectric constant oxides. *Eur. Phys. J. Appl. Phys.* **2004**, *28*, 265–291. [CrossRef]
24. Higashi, G.S.; Fleming, C.G. Sequential surface chemical reaction limited growth of high quality Al_2O_3 dielectrics. *Appl. Phys. Lett.* **1989**, *55*, 1963–1965. [CrossRef]

25. Soto, C.; Tysoe, W.T. The reaction pathway for the growth of alumina on high surface area alumina and in ultrahigh vacuum by a reaction between trimethyl aluminum and water. *J. Vac. Sci. Technol. A Vac. Surf. Films* **1991**, *9*, 2686–2695. [CrossRef]
26. Masuda, H.; Fukuda, K. Ordered metal nanohole arrays made by a two-step replication of honeycomb structures of anodic alumina. *Science* **1995**, *268*, 1466–1468. [CrossRef] [PubMed]
27. Prida, V.M.; Sanz, R.; Vega, V.; Navas, D.; Pirota, K.R.; Asenjo, A. Self-assembled nanoporous oxide membranes. In *Encyclopedia Nanosci. Nanotechnol*; Nalwa, H.S., Ed.; American Scientific Publishers: Valencia, CA, USA, 2011; Volume 22, pp. 509–532.
28. Geyer, S.M.; Methaapanon, R.; Johnson, R.; Brennan, S.; Toney, M.F.; Clemens, B.; Bent, S. Structural evolution of platinum thin films grown by atomic layer deposition. *J. Appl. Phys.* **2014**, *116*, 064905. [CrossRef]
29. Banerjee, P.; Lee, W.-J.; Bae, K.-R.; Lee, S.B.; Rubloff, G.W. Structural, electrical, and optical properties of atomic layer deposition Al-doped ZnO films. *J. Appl. Phys.* **2010**, *108*, 043504. [CrossRef]
30. Hou, Q.; Meng, F.; Sun, J. Electrical and optical properties of Al-doped ZnO and $ZnAl_2O_4$ films prepared by atomic layer deposition. *Nanoscale Res. Lett.* **2013**, *8*, 144. [CrossRef] [PubMed]
31. Elam, J.W.; George, S.M. Growth of ZnO/Al_2O_3 alloy films using atomic layer deposition techniques. *Chem. Mater.* **2003**, *15*, 1020–1028. [CrossRef]
32. Meng, X.; Banis, M.N.; Geng, D.; Li, X.; Zhang, Y.; Li, R.; Abou-Rachid, H.; Sun, X. Controllable atomic layer deposition of one-dimensional nanotubular TiO_2. *Appl. Surf. Sci.* **2013**, *266*, 132–140. [CrossRef]
33. Bachmann, J.; Zierold, R.; Chong, Y.T.; Hauert, R.; Sturm, C.; Schmidt-Grund, R.; Rheinländer, B.; Grundmann, M.; Gösele, U.; Nielsch, K. A practical, self-catalytic, atomic layer deposition of silicon dioxide. *Angew. Chem. Int. Ed.* **2008**, *47*, 6177–6179. [CrossRef] [PubMed]
34. Król, K.; Sochacki, M.; Taube, A.; Kwietniewski, N.; Gierałtowska, S.; Wachnicki, Ł.; Godlewski, M.; Szmidt, J. Influence of atomic layer deposition temperature on the electrical properties of $Al/ZrO_2/SiO_2$/4H-SiC metal-oxide semiconductor structures. *Phys. Status Solidi A* **2018**, *215*, 1700882. [CrossRef]
35. Uribe-Vargas, H.; Molina-Reyes, J.; Romero-Morán, A.; Ortega, E.; Ponce, A. Gate modeling of metal–insulator–semiconductor devices based on ultra-thin atomic-layer deposited TiO_2. *J. Mater. Sci. Mater. Electron.* **2018**, *29*, 15761–15769. [CrossRef]

© 2018 by the authors. Licensee MDPI, Basel, Switzerland. This article is an open access article distributed under the terms and conditions of the Creative Commons Attribution (CC BY) license (http://creativecommons.org/licenses/by/4.0/).

Review

Metal Fluorides as Lithium-Ion Battery Materials: An Atomic Layer Deposition Perspective

Miia Mäntymäki [1,2,*], Mikko Ritala [2] and Markku Leskelä [2]

1. Picosun Oy, Tietotie 3, FI-02150 Espoo, Finland
2. Department of Chemistry, P.O. Box 55 (A. I. Virtasen aukio 1), University of Helsinki, FI-00014 Helsinki, Finland; mikko.ritala@helsinki.fi (M.R.); markku.leskela@helsinki.fi (M.L.)
* Correspondence: miia.mantymaki@picosun.com; Tel.: +358-50-320-6435

Received: 9 June 2018; Accepted: 6 August 2018; Published: 8 August 2018

Abstract: Lithium-ion batteries are the enabling technology for a variety of modern day devices, including cell phones, laptops and electric vehicles. To answer the energy and voltage demands of future applications, further materials engineering of the battery components is necessary. To that end, metal fluorides could provide interesting new conversion cathode and solid electrolyte materials for future batteries. To be applicable in thin film batteries, metal fluorides should be deposited with a method providing a high level of control over uniformity and conformality on various substrate materials and geometries. Atomic layer deposition (ALD), a method widely used in microelectronics, offers unrivalled film uniformity and conformality, in conjunction with strict control of film composition. In this review, the basics of lithium-ion batteries are shortly introduced, followed by a discussion of metal fluorides as potential lithium-ion battery materials. The basics of ALD are then covered, followed by a review of some conventional lithium-ion battery materials that have been deposited by ALD. Finally, metal fluoride ALD processes reported in the literature are comprehensively reviewed. It is clear that more research on the ALD of fluorides is needed, especially transition metal fluorides, to expand the number of potential battery materials available.

Keywords: atomic layer deposition (ALD); lithium-ion batteries; fluoride; thin films

---

1. Introduction

The technological advancements that have taken place during the last few decades have created the need to store more energy in ever smaller volumes. Lithium-ion batteries can store large amounts of energy in small weights and volumes, making them the technology-of-choice for multiple applications. In addition to their use in everyday portable electronics and all-electric vehicles, lithium-ion batteries could also provide a way to store large amounts of energy harnessed by using solar cells and wind turbines [1]. It is clear that the market for lithium-ion batteries will continue to increase in the years to come [2,3]. However, urgent materials engineering advances are needed to enable the continued progress of the technology and answer the high demands of future applications.

The most widely used lithium-ion battery materials include oxides and phosphates for cathodes [4,5]. These materials are intercalation electrodes and thus have relatively low usable capacities of 100–150 mAh/g. Graphitic carbon is the most frequently used anode material, with spinel lithium titanate and elemental silicon gaining more interest: the titanate for its enhanced safety and elemental silicon for its huge maximum capacity of 3600 mAh/g, made possible by its alloying reaction with lithium [4,5]. Alloying anodes have presented challenges in battery fabrication, because large volume changes associated with the alloying reactions can lead to electrode pulverization and degradation. However, with recent advances in nanofabrication, these problems might soon be mitigated by the use of proper micro- and nanoscale battery constructions [6–10].

Whereas high capacity alternatives, such as silicon, are already being intensely studied to replace the present anode materials, oxide and phosphate materials still dominate the research on cathodes, despite their limited capacities. Fluorides were studied extensively in the 1960s and 1970s for use in primary lithium batteries (i.e., batteries that would not be rechargeable) due to their high theoretical capacities and energy densities [11]. It was hoped that these materials would act as conversion cathodes, forming lithium fluoride during discharge (Equation (1)):

$$n\text{Li}^+ + ne^- + \text{MF}_n \rightarrow n\text{LiF} + \text{M} \qquad (1)$$

Conversion cathodes generally suffer from the same volume change problems as alloying anodes. However, by using the same nanofabrication methods as for anodes, these materials could provide a new, interesting class of lithium-ion battery electrode materials. Additionally, fluorides have also been reported to show high lithium-ion conductivities, making them possible solid electrolytes for all-solid-state thin film batteries [12–14].

Many enhancements in lithium-ion battery properties can be achieved by depositing component thin films onto 3D substrate geometries, as both active materials and protective layers on the active materials. During the 21st century, the thin film manufacturing method known as atomic layer deposition (ALD) has become a vital enabler of progressive technology nodes in microelectronics and is now becoming similarly important in the fabrication of Li-ion batteries. The advantages of ALD, including high film uniformity and excellent conformality over high-aspect-ratio substrates, make it ideal for the deposition of materials for ever-smaller, more complicated batteries: strict conformality is especially important for integrated, all-solid-state batteries, in which the small electrode thin film thicknesses can still produce high energy densities per footprint area when deposited into deep trenches [8]. The deposition of solid electrolyte materials, instead of using the current liquid electrolytes, could solve many of the safety issues associated with lithium-ion batteries.

This review briefly introduces the basic concept of lithium-ion batteries, some of the materials currently used in these batteries and the use of ALD in depositing these materials. Liquid electrolytes will not be discussed but instead potential inorganic solid electrolytes are shortly reviewed. Fluoride materials are presented as a potential "new" class of battery materials with uses as both electrodes and solid electrolytes for lithium-ion batteries. To motivate further studies on fluoride deposition using ALD, the literature in this area is reviewed.

2. The Lithium-Ion Battery

2.1. Basic Principle

Lithium-ion batteries are used for energy storage in applications ranging from cell phones and laptops to electric vehicles. The basic concept of a lithium-ion battery is the same as for any other battery: chemical energy stored in the electrodes is converted into electrical energy via a chemical reaction. Some of the battery types in use today are depicted in Figure 1. As can be seen, lithium-ion batteries have surpassed many of the older battery technologies in energy density. This is related to the fact that lithium ions are relatively small and light-weight, which makes it possible to obtain high energy densities from lithium-containing materials. In addition, lithium forms compounds with large enough potentials to produce so-called "high quality" energy, or high power densities [15]. The small size of the lithium-ion is also an advantage in that lithium-ions are highly mobile in many materials, ensuring only low energy losses due to kinetic effects.

Figure 2 depicts the basic schematic of a lithium-ion battery. The battery consists of two electrodes (the positive cathode and the negative anode), an electrolyte and two current collectors. Because the active materials in the battery, the electrodes, are separated by an inert electrolyte, the energy liberated in the chemical reaction between the electrode materials can be converted into electrical work. During battery operation, lithium-ions flow through the electrolyte from the anode to the cathode.

At the same time electrons flow in the outer circuit from the anode to the cathode, balancing the net charge flow. These electrons can be used to do work, such as operate a laptop.

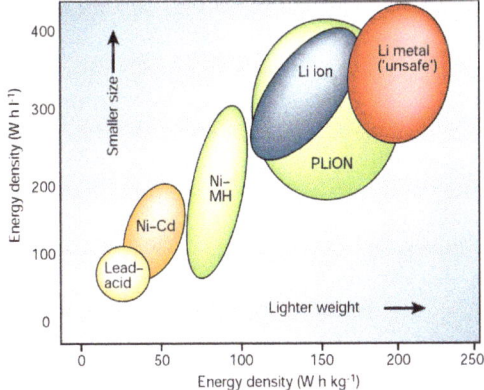

Figure 1. Comparison of energy densities in different battery types in use today. Reprinted with permission from [16]. Copyright 2001 Springer Nature.

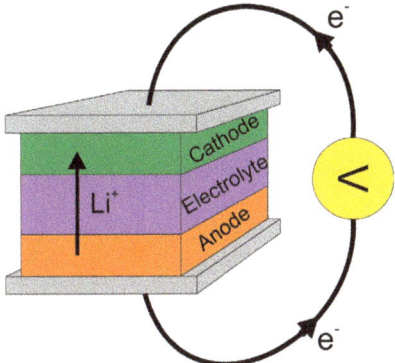

Figure 2. Basic schematic of a lithium-ion battery, showing the flow of lithium ions and electrons during discharge.

In addition to the high energy and power densities, interest in lithium-ion batteries derives from the fact that most of them can be recharged multiple times without prohibitive loss of battery capacity. During recharging a voltage is applied to the outer circuit, which reverses the flow of electrons and Li^+-ions, moving them from the cathode to the anode. During the charge-discharge cycling, the battery capacity should stay as constant as possible. Usually the capacity slowly degrades with increased cycling as a result of, for example, reactions between the electrolyte and the electrodes and changes in the morphology of the electrodes, such as breakage or delamination [15].

Most lithium-ion batteries employ an organic liquid electrolyte material, composed of organic carbonates such as ethylene, dimethyl and diethyl carbonates, lithium hexafluorophosphate and potentially different additives [5,15]. Liquid electrolytes enable very high lithium-ion conductivities, which are beneficial for the battery operation. At the same time, the electrolyte should be an electron insulator so that no self-discharge takes place [15]. Liquid electrolytes can accommodate volume changes in the electrodes during cycling, which reduces stress in the battery. However, liquid electrolytes are also a major reason for the safety concerns of lithium-ion batteries: the electrolyte can decompose at the electrodes, most often on the anode and form a solid-electrolyte interphase layer, or SEI-layer [15]. Commonly the SEI-layer protects the electrodes from further reactions with the electrolyte but in the event of an incomplete SEI-formation, reactions can proceed further with pressure building up inside the battery. Since the electrolyte materials are flammable, explosions can occur [17].

2.2. Conventional Electrode Materials

The electrodes define the capacity and voltage of any battery. Since both lithium-ions and electrons move inside the electrodes, the materials need to be both ion and electron conductors. The reactions taking place in the electrodes can be roughly divided into conversion reactions, alloying reactions and insertion or intercalation reactions (Figure 3) [4]. Conversion and alloying electrodes generally provide higher battery capacities but there can be problems with electrode degradation during cycling due to repeated volume changes associated with the conversion reaction. Thus, conversion and alloying electrodes have more often been used in primary batteries, meaning non-rechargeable batteries.

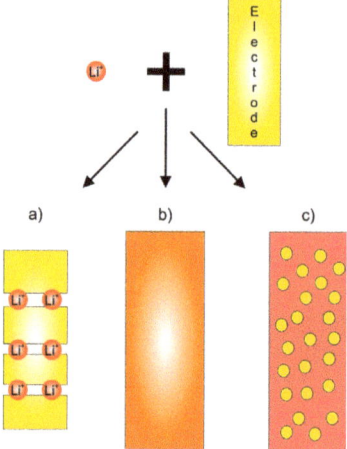

Figure 3. Schematic illustrating the different kinds of reactions possible with lithium ions and electrode materials. (**a**) Intercalation; (**b**) Alloying; (**c**) Conversion. With alloying and conversion reactions, significant volume changes often take place in the electrode.

In rechargeable, or secondary, batteries intercalation reactions are the most commonly used, with the $LiCoO_2$–C_6 (graphite) battery being a prime example of this reaction. Even if intercalation electrodes do not generally suffer from volume changes of similar magnitude to conversion electrodes, they still require care in the charging-discharging process: too deep charge-discharge cycling can lead to irreversible changes in the electrode structure, which in turn can lead to a diminished battery capacity.

2.2.1. Cathodes

Intercalation cathode materials are most often lithium containing transition metal oxides, in which the insertion and extraction of lithium is made possible by the redox capabilities of the transition metal [4,18]. Although $LiCoO_2$ is the most commonly used cathode material, manganese and nickel oxides are gaining interest as possible future cathodes [18,19]. Table 1 contains the potentials and capacities obtainable from some of the most studied cathode materials. Despite their common use in batteries today, transition metal oxide cathodes have some serious drawbacks. Firstly, the potentials obtainable from these materials are limited, which is a problem especially for high power density applications. Secondly, the capacities available from these materials are quite low as a result of the limited amount of lithium available before irreversible changes in the structures occur. For example, only half of the lithium-ions in LiCoO$_2$ can be reversibly utilized [19]. Lastly, liquid electrolytes are known to give rise to dissolution of the transition metal, thus decreasing the cathode capacity even further [18]. However, thin film methods such as atomic layer deposition can be used to deposit thin layers (e.g., Al$_2$O$_3$ or AlF$_3$) onto cathode materials to protect them from side reactions such as transition metal dissolution [20–24]. Some of these problems can also be circumvented by using mixture cathodes, such as LiNi$_{(1-y-z)}$Mn$_y$Co$_z$O$_2$, which can also produce slightly higher capacities [5,18,19]. In addition to oxides, sulphates and phosphates, such as LiFePO$_4$, have also been studied extensively [19]. Despite being cheaper and less toxic than most cathode materials and providing a reasonable capacity, LiFePO$_4$ suffers from low electronic conductivity, limiting its applicability [19]. For cathodes, alloying and conversion reactions have not gained much attention.

Table 1. Potentials and capacities obtainable from some of the most studied lithium-ion battery cathode materials.

Material	Type	Potential (V) (vs. Li/Li$^+$)	Specific Capacity (mAh/g)	Ref.
LiCoO$_2$	Intercalation	~3.9	140	[4]
LiNiO$_2$	Intercalation	2.7–4.1	140–200	[4,19]
LiMn$_2$O$_4$	Intercalation	3.5–4.5	150	[4]
LiFePO$_4$	Intercalation	3.4	170	[4,5,19]
V$_2$O$_5$	Intercalation	3.2–3.4	120	[4]

2.2.2. Anodes

Table 2 presents some potential anode materials for lithium-ion batteries. For an anode, a potential as low as possible is desired to reach a high cell voltage. As with the cathode materials, intercalation anodes are more common than the other types of anodes. However, a lot of work is now being put into studying materials such as silicon and tin as lithium-ion battery alloying anodes [4,5,9,25].

Table 2. Potentials and capacities obtainable from some of the most studied lithium-ion battery anode materials.

Material	Type	Potential (V) (vs. Li$^+$/Li)	Specific Capacity (mAh/g)	Ref.
C$_6$	Intercalation	<0.6	370	[4,5]
Li$_4$Ti$_5$O$_{12}$	Intercalation	1.5	175	[4]
Li	Alloying	0	3800	[26]
Si	Alloying	0.1–0.3	3580	[4,5]
Sn	Alloying	0.6–0.8	990	[4,5,9]
MO (M = Co^{2+}, Fe^{2+}, Cu^{2+}, Ni^{2+})	Conversion	~1.8–2.0	670–750	[27–29]

Metallic lithium was, unsurprisingly, the first choice for the anode of a lithium-ion battery [18,26]. However, safety concerns such as dendrite formation [26] have moved the interest towards other anodes such as graphitic carbon C$_6$, which produces capacities of 370 mAh/g [5,15]. The use of C$_6$ still does not help avoid dendritic lithium deposition due to the very low electrode potential of carbon anodes. One possible replacement for carbon is the spinel lithium titanate Li$_4$Ti$_5$O$_{12}$, which has a reasonable capacity with a very small volume expansion during lithiation. In addition, the titanate is environmentally benign with a reasonable cost [28]. A major drawback of this material and many insertion oxide anodes in general, is the high electrode potential, which results in cells with low operation voltages [5,28].

Of the alloying anodes, elemental silicon has attracted much attention because of its low cost, abundance in nature and high specific capacity of over 3500 mAh/g [28]. However, alloying the maximum 3.75 lithium-ions per one Si atom produces volume changes of up to 300% in the electrode, limiting the cycling ability of the anode [5]. Nanostructured silicon is now studied in the hope of alleviating the problems with volume expansion [5,9]. In addition, atomic layer deposition has been utilized in contact with Si anodes to improve their mechanical integrity and to stabilize the interfaces of the anode [10].

Similarly, elemental tin has attracted much attention but suffers from the same problems as silicon [9]. Attempts have been made to circumvent the volume change problems by using conversion anodes composed of, for example SnO or SnO_2. In these anodes, lithium first forms lithium oxide and tin is reduced to metallic tin, which can further alloy lithium and is responsible for the reversible capacity [28]. However, the problems related to volume changes still persist to some extent in these anodes [28]. Many transition metal oxides have also been studied as conversion anodes, with capacities ranging from 600 to 700 mAh/g [27]. In addition to the persisting volume change problem, conversion anodes often also suffer from large overpotentials [9].

2.3. Conventional Solid Electrolyte Materials

Solid electrolyte materials are the enablers of all-solid-state Li-ion batteries. These materials have stringent property demands: they must be unreactive at the electrode potentials and have a high lithium-ion conductivity at room temperature [15,30]. In addition, they must be good electron insulators to avoid self-discharge and short-circuits [4]. Despite the high specification for material properties, a vast number of solid electrolyte materials suitable for lithium-ion batteries have been reported in the literature [30–33]. In addition to inorganic ceramics, composites and polymer mixtures can be used as solid electrolytes [31,33,34].

Inorganic fast lithium-ion conducting materials can be single-crystalline, polycrystalline or amorphous, with many different structural types reported for the crystalline materials (Table 3, Figure 4) [31,32]. Generally, an amorphous electrolyte material would be preferred as grain boundaries in crystalline materials can lead to both impeded ion movement and electron leakage and thus poorer insulating properties [33,34]. In addition, amorphous materials can provide isotropic lithium-ion conduction in a wide range of different compositions [32].

Table 3. Structures and ionic conductivities of some of the most studied solid inorganic lithium-ion conducting materials.

Material	Structure	Ionic Conductivity at RT (S/cm)	Ref.
LiPON	Amorphous	10^{-8}–10^{-6}	[31,32]
Li_2O–SiO_2–V_2O_5 (LVSO)	Crystalline/Amorphous	10^{-7}–10^{-5}	[32]
Li_2S–GeS_2–Ga_2S_3	Amorphous	10^{-4}	[32]
(Li,La)TiO_3 (LLT)	Perovskite	10^{-3}	[31,33]
$LiTi_2(PO_4)_3$	NASICON	10^{-5}	[31,32]
$Li_{14}ZnGe_4O_{16}$	LISICON	10^{-6}	[33]
$Li_6BaLa_2Ta_2O_{12}$	Garnet	10^{-5}	[31]

Amorphous, or glassy, electrolytes can be roughly divided into oxide and sulphide glasses [32]. Probably the most studied amorphous solid electrolyte material is nitrogen doped lithium phosphate (lithium phosphorus oxynitride), LiPON, which is in fact already in use in thin film lithium-ion batteries [32,33]. The material is usually deposited by sputtering in a nitrogen atmosphere and the resulting films have lithium-ion conductivities of the order of 10^{-8}–10^{-6} S/cm [32], as compared to the ~10^{-2} S/cm for liquid electrolytes [5]. The success of LiPON comes not only from its reasonably high lithium-ion conductivity but also from its excellent stability against the electrode materials [32]. Sulphide glasses, on the other hand, have been studied much less than the corresponding oxides, mostly because of their reactivity in air and corrosiveness [32].

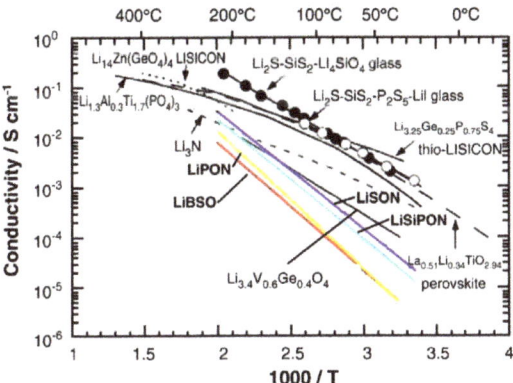

Figure 4. Bulk ionic conductivities of selected oxides and phosphates as a function of temperature. Reprinted with permission from [33]. Copyright 2009 Elsevier.

Out of crystalline electrolyte materials, the perovskite (Li,La)TiO$_3$ (LLT) is known to show high bulk conductivity. However, reduction of Ti^{4+} and the consequent increase in electronic conduction are problems associated with this material [31,33]. Titanium-free perovskites, such as (Li,La)NbO$_3$ have thus been studied [32]. Phosphates with the NASICON (sodium superionic conductor) structure are also known to show high lithium-ion conductivities [32]. In Table 3 LiTi$_2$(PO$_4$)$_3$ is given as an example but many other +IV oxidation state metals can be substituted for titanium in the structure. LISICON (lithium superionic conductor) materials, on the other hand, are mixtures of lithium silicates or germanates, lithium phosphates or vanadates and lithium sulphates and can produce similar conductivities as the NASICON structures [32]. In addition to these material classes, oxides such as Li$_6$BaLa$_2$Ta$_2$O$_{12}$ with the garnet structure have been studied extensively and shown to have reasonable ionic conductivities [31]. Another benefit of the garnet materials is their high chemical stability in contact with the electrodes [32].

3. Metal Fluorides as Lithium-Ion Battery Materials

Metal fluorides can be utilized in lithium-ion batteries in many ways [11]. This review will focus only on metal fluorides as electrodes, artificial SEI-layers and solid electrolytes. The reader is advised that much work has also been done on other fluorinated materials in batteries, such as fluorinated salts as additives in liquid electrolytes [11,35], fluorinated solvents in batteries [35], carbon fluorides as negative electrodes [11] and fluorinated binder materials [11,35]. These topics will not be discussed further here.

3.1. Electrode Materials

Amatucci and Pereira note in their review on metal fluoride based electrode materials that "*The use of fluorides stems from the intrinsic stability of fluorinated materials and their ability to generate high electrochemical energy as electrodes*" [11]. Indeed, metal fluoride cathodes generally produce higher potentials than the corresponding oxides of the same redox-couple, which can lead to higher energy densities [11]. Thus, fluoride materials could be used in high voltage batteries, where the stability of active materials is especially important. Fluorides can be used as cathodes either as pure fluorides or as doped materials, such as oxyfluorides, fluorosulphates or fluorophosphates [11,35]. Fluoride doping has been reported to improve capacity retention of intercalation cathodes such as lithium nickel oxide and lithium nickel cobalt oxide [11]. This could be related to a slower dissolution of transition metals into liquid electrolytes from the oxyfluorides [11]. For fluorophosphate cathodes, such as Li$_2$CoPO$_4$F, high potentials of over 5 V are obtainable, accompanied again by a slower dissolution of the transition metal [11,35]. However, the performance of these cathodes is limited due to poor ionic and electronic conductivity and instability of liquid electrolytes at such high potentials [35].

As already mentioned in the Introduction, pure metal fluorides gained interest decades ago as electrodes for primary batteries because of their high capacities (Figure 5). With the increased interest in high capacity alloying and conversion anodes such as Si and SnO_2, fluoride conversion cathodes could also resurface as interesting materials for secondary lithium-ion batteries [11,25,29,36–39]. Similar to alloying anodes, fluoride conversion cathodes suffer from large volume changes and subsequent pulverization during cycling. In addition, fluorides are very poor electron conductors due to their high band gaps and often show high overpotentials, which can make their integration as reversible electrodes challenging [25,29]. In the early years materials such as CuF_2 and HgF_2 were studied but with little success. Recently, interesting progress has been made in this research area. For example, BiF_3, FeF_2 and FeF_3 have been studied extensively as cathodes using nanocomposites of the metal fluorides and conductive carbon [38,40]. Using nanocomposites with carbon can not only help with the volume change but also with the inherently low electrical conductivity of fluoride materials.

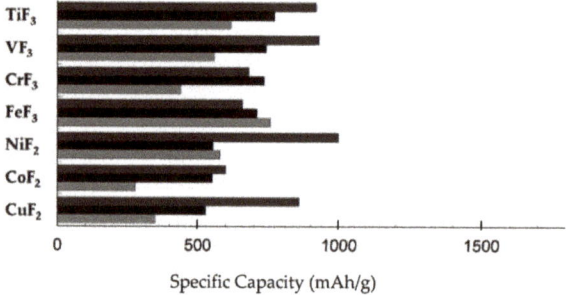

Figure 5. Theoretical (black), first discharge (dark grey) and charge (light grey) specific capacities of conversion fluoride cathode materials. Adapted with permission from [36]. Copyright 2010 Wiley-VCH.

Due to the work on primary batteries pure fluorides are generally considered only as conversion cathodes but some reports on intercalation fluorides have been published [41–43]. For example, Li_3FeF_6 has been reported to show intercalation of 0.7–1 Li^+ ions per fluoride unit in a carbon nanocomposite form, resulting in a reversible capacity of 100–140 mAh/g [41,42]. The capacity depends on the size of the Li_3FeF_6 particles, with smaller particles resulting in a higher capacity [42]. A deeper discharge of the material was reported to lead to LiF formation, indicating a conversion reaction at low potentials. Similarly, a nanocomposite of Li_3VF_6 was reported to reversibly intercalate up to one Li^+ per fluoride unit [43]. Calculations predict that fluorides such as $LiCaCoF_6$ could provide very high intercalation voltages [44].

In addition to their potential use as electrode materials, fluorides can also be utilized as solid-electrolyte-interface layers deposited on the more conventional electrode materials to protect them from reactions with the organic liquid electrolytes. AlF_3 has been studied extensively in this regard, on both cathodes [45–49] and anodes [50]. AlF_3 is suitable for electrode protection because it is rather inert and the Al^{3+}-ion cannot be reduced or oxidized in battery conditions [20]. The material has been reported to decrease the irreversible capacity losses of electrodes and improve cycling stability, [45,47] and increase the thermal stability of electrodes [45,46,49]. Figure 6 illustrates how a layer of AlF_3 can increase the capacity retention in a lithium cobalt nickel manganese oxide cathode. Using too much AlF_3, however, decreases the capacity considerably.

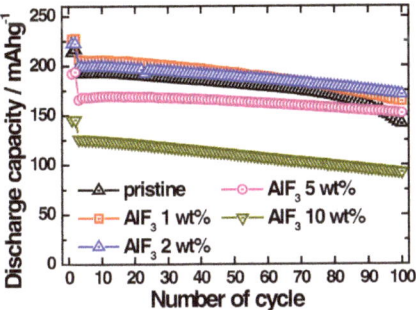

Figure 6. The effect of an AlF$_3$ coating on the discharge capacity of a lithium cobalt nickel manganese oxide cathode as a function of the number of charge/discharge cycles. Reprinted with permission from [47]. Copyright 2012 Wiley-VCH.

3.2. Solid Electrolyte Materials

The applicability of metal fluorides as solid electrolytes for Li$^+$ ions has not been studied as extensively as their use as electrodes. However, some examples of potential electrolyte materials can be found in the literature. Li$_3$AlF$_6$, a stoichiometric ternary of LiF and AlF$_3$, has been reported to show high ionic conductivities of the order of 10^{-6} S/cm in thin film form [12–14,51]. In addition, milling this ternary fluoride with LiCl has been reported to lead to high conductivities [52]. Other fluorides that also show high conductivities when mixed with LiF include NiF$_2$, VF$_3$, CrF$_3$ and YF$_3$ (Figure 7) [13,53]. These materials have been deposited by thermal evaporation and fast quenching, resulting in amorphous thin films. The increased ionic conductivity in these mixtures is attributed to the formation of amorphous intermediate phases with high coordination numbers for lithium, such as in the Li$_3$AlF$_6$ phase [13]. Even more complicated fluoride mixtures have been studied as well [12,14,54–56], such as the LiF–AlF$_3$–ScF$_3$ system, which can reach similar conductivity values as the pure Li$_3$AlF$_6$ [12]. With fluoride glasses of the type LiF–ZrF$_4$–LaF$_3$ high lithium-ion conductivities can be obtained for materials with sufficient LiF component [54].

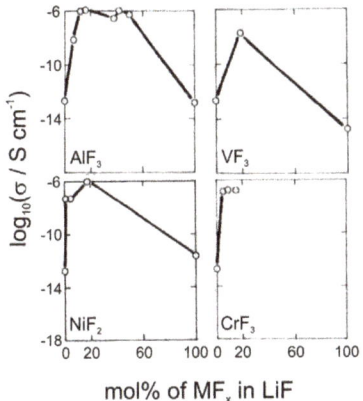

Figure 7. Room temperature ionic conductivities log$_{10}\sigma$ of fluoride thin films composed of LiF and AlF$_3$, VF$_3$, NiF$_2$ or CrF$_3$. Adapted with permission from [13]. Copyright 1984 Elsevier.

In addition to the applications in lithium-ion batteries, some metal fluoride mixtures can act as electrolytes for F$^-$-ions, making high voltage fluoride-ion batteries a possibility [57–61]. Mixed fluoride glasses can, in some cases, conduct both lithium- and fluoride-ions, depending on the molar ratios of the metal fluorides [54,56].

4. Atomic Layer Deposition

4.1. Basic Principle

Atomic layer deposition (ALD) is a gas phase thin film deposition method, best known for producing thin films of high uniformity and conformality. It is closely related to chemical vapour deposition (CVD). Whereas in CVD gaseous precursors are supplied simultaneously, in ALD precursor pulses are separated by purge gas pulses or evacuation periods, resulting in no gas phase reactions. Instead, the precursors react one at a time with the substrate or film surface groups in a digital manner [62,63]. ALD has different variations, including thermal ALD [64], plasma-enhanced ALD (PEALD) [65] and photo-ALD [66,67], depending on how energy is supplied to the deposition reaction. Thermal ALD refers to a process where the energy for the surface reactions is produced by heating. In PEALD, additional energy from radicals and, depending on the reactor configuration, possibly also ions and electrons, is used [65]. In photo-ALD reactions are enhanced with UV-and/or visible light [67].

The atomic layer deposition cycle is composed of four steps (Figure 8). In step 1, the first precursor adsorbs and reacts on a substrate surface. After all potential surface sites have reacted with the first precursor, excess precursor molecules and side products are purged or pumped away in step 2. In step 3, the second precursor reacts with the surface, forming a binary film. In step 4 reaction by-products and excess precursor two are purged and pumped away. By repeating the four-step cycle, a film of desired thickness can be formed [62]. Generally, a film of one monolayer or less is formed in one ALD cycle [68]. The amount of material deposited depends both on the density of active surface groups and the size of the precursor molecules [62,68].

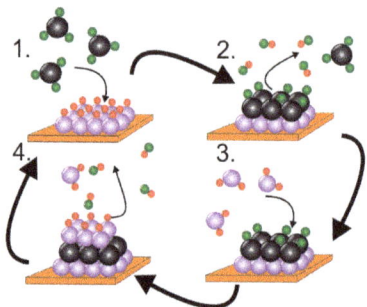

Figure 8. Schematic illustrating one cycle in ALD. 1. Precursor 1 molecules react with a surface covered with active sites; 2. Reaction side products and excess precursor are purged away; 3. Precursor 2 is introduced and it reacts with the surface covered with precursor 1 molecules; 4. The reaction side products and excess of precursor 2 are again purged away, resulting ideally in one monolayer of material on the substrate.

In ALD literature, the reaction type illustrated in Figure 8 has been traditionally called "ligand exchange" [68]: for example, in the case of Al_2O_3 deposition trimethylaluminum (TMA) and water react in a way that methane is produced as a side product. Thus, it can be viewed as methyl ligands changing their bonding from aluminum in TMA to hydrogen from hydroxide surface groups. In synthesis work, this type of reaction is commonly called metathesis. This broad definition of ligand exchange can be applied to most ALD reactions in use today. Other ALD-type reactions include combustion with ozone and oxygen radicals, an additive reaction with elemental precursors and controlled decomposition of an adsorbed species [64]. Ideal ALD processes should show an ALD window, meaning a temperature region where the film growth rate stays constant as a function of deposition temperature. However, usually small deviations from this behaviour are seen, because the number of active surface sites can be highly dependent on the deposition temperature [64]. As most often less than a monolayer is deposited per one ALD cycle, the slowness of the deposition process is considered as one of the greatest weaknesses of ALD [62].

All reactions in ALD occur between surface groups and adsorbing gaseous precursors, so the reactions become terminated when all the surface groups have reacted or when the steric hindrance from large precursor molecules prevents further precursor adsorption [68]. This results in self-limiting growth which means that using higher precursor doses, often in practice meaning longer precursor pulse times, will not result in more growth and that a constant amount of film is deposited in each cycle [62,64]. To achieve this self-limiting or saturative behaviour, it is important that the precursor does not self-decompose. In addition, long enough purge times between precursor pulses are required to ensure that no excess precursor remains in the gas phase or adsorbed onto the surface when the second precursor is introduced. Self-limiting growth enables the large area uniformity, excellent conformality and nanometre-level thickness control of ALD films: the growth per cycle is constant, which means that the film thickness can be specified by choosing a proper cycle number. Thus, with ALD even very demanding 3D structures can be covered with a film of constant thickness, when long enough precursor pulse times and purges are employed [62].

ALD is a promising method for the deposition of small, integrated 3D all-solid-state batteries due to the precision it affords in thickness control and conformality. All-solid-state batteries can be integrated into, for example, microelectromechanical systems (MEMS) to achieve autonomous sensing devices. For this type of integration, very small batteries are generally required because the size of the battery can limit the size of the whole device [8,17]. Making a battery smaller by using thinner active layers is a viable solution for all-solid-state batteries because thinner layers result in smaller transport losses and over-potentials due to smaller diffusion length scales [8,17,69]. Importantly, the limitations imposed by the low lithium-ion conductivities of solid electrolytes can thereby be circumvented [8]. However, thin electrode layers limit the energy available from the battery. By making batteries smaller with complex 3D structures, gains in both energy and power density can be achieved due to simple geometrical reasons: more active material can be packed into a smaller foot print area, with the advantages of short diffusion lengths still present [8].

4.2. Atomic Layer Deposition of Conventional Lithium-Ion Battery Materials

In this section, a few examples of the more conventional lithium-ion battery materials deposited by ALD are introduced. Examples of binary oxide electrodes (specifically V_2O_5 and TiO_2) and lithium containing materials are presented. Lithium containing materials are quite a new addition to the ALD materials toolbox—the first paper on the subject was published in 2009 [70]. Since then, this area of research has expanded very rapidly. Due to the mobility and reactivity of Li^+ ions, ALD of lithium containing materials has additional process development issues in comparison to most other ALD processes and some of these issues are discussed in the following subsections. In this section emphasis is given to potential solid electrolyte materials, as these are generally considered the most difficult materials to deposit. For a more thorough review of this subject, both review articles and books are available [71–79]. In recent years, ALD has also been studied extensively as a method to modify the interfaces between the electrodes and the electrolyte by forming an artificial SEI-layer [75,80,81]. ALD Al_2O_3 is generally used for this application and it has been found that a few ALD cycles can improve the cycling capability and capacity retention of the electrodes [80,81]. These results will not be discussed further here but the reader is advised that a lot of literature on this subject is available [80–83]. Some examples of ALD-made metal fluorides as artificial SEI layers will be shortly mentioned in Section 5.

4.2.1. Cathodes

Table 4 includes examples of some of the conventional lithium-ion battery cathode materials deposited by ALD. Both binary and ternary oxides and phosphates have been deposited for conventional batteries. Li_2S has been envisioned as a cathode for lithium-sulphur batteries [84].

Table 4. Examples of conventional lithium-ion battery cathode materials deposited by ALD. Abbreviations used: O^iPr = iso-propoxide, O^tBu = tert-butoxide, Cp = cyclopentadienyl, TMPO = trimethyl phosphate, thd = 2,2,6,6-tetramethyl-3,5-heptanedionato.

Material	Precursors	T_{Dep} (°C)	Growth Rate (Å)/Binary Cycle	Capacity (mAh/g)	Ref.
V_2O_5	$VO(O^iPr)_3 + H_2O$	105	0.15–0.2	455	[85,86]
V_2O_5	$VO(O^iPr)_3 + O_3$	170–185	0.25	440	[86]
Li_xCo_yO	$LiO^tBu + CoCp_2 + O_2$ plasma	325	1.0 with Li:Co pulsing ratio 1:1	96 for Li:Co pulsing 1:4	[87,88]
$Li_xFe_yPO_4$	$LiO^tBu + FeCp_2 + O_3 +$ $TMPO + H_2O$	300	0.85 with Li:Fe pulsing ratio 1:5	150 at 0.1 C	[89]
Li_xMn_yO	$Lithd + Mn(thd)_3 + O_3$	225	0.2 with Li:Mn pulsing ratio 1:19	–	[90]
Li_xMn_yO	$LiO^tBu + H_2O$ or $Lithd + O_3$ on MnO_2	225	–	230	[90]
Li_2S	$LiO^tBu + H_2S$	150–300	1.1 at 150–300 °C	500	[91]

Vanadium oxide was one of the earliest materials to be deposited by ALD for lithium-ion batteries [85,86,92]. Vanadyl tris-iso-propoxide with either water or ozone as co-reactant have been used as precursors for this material. Using water produces amorphous films, while with ozone crystalline films can be obtained. Crystalline and amorphous films produce different capacities depending on the extent of lithium intercalation, with crystalline films having higher capacities when 1 or 2 Li-ions intercalate per one V_2O_5 unit [86]. A surprisingly high capacity of 455 mAh/g has been reported for 200 nm of amorphous V_2O_5 between 1.5 and 4.0 V (Li/Li$^+$) [85]. This high value is related to the large potential range used for cycling, resulting in 3 Li-ions intercalating per one V_2O_5 [85]. For a thicker amorphous film of 450 nm, a capacity of 275 mAh/g was obtained in the same range. Both the thicker and the thinner films showed reasonable capacity retention after 90 cycles. For crystalline films, capacities of 127–142 mAh/g have been obtained in the potential range 2.6–4.0 (1 Li per V_2O_5) [86]. Between 1.5 and 4.0 V (3 Li per V_2O_5), a capacity of 440 mAh/g is obtainable but the capacity degrades to 389 mAh/g already in the second cycle.

As already discussed, lithium cobalt oxide $LiCoO_2$ is currently the most often used cathode material in lithium-ion batteries. Despite this, only two reports from one group on the deposition of $LiCoO_2$ by ALD can be found in the literature [87,88]. It appears that the challenges in cobalt oxide deposition have had an effect on the research of the lithiated material. The reported $LiCoO_2$ process makes use of oxygen plasma combined with $CoCp_2$ (cobaltocene) and LiO^tBu (lithium tert-butoxide). The deposition supercycle consists of Co_3O_4 and Li_2CO_3 subcycles and the effect of different pulsing ratios on film properties was studied. The process showed saturation with both metal precursors with Li:Co = 1:1 pulsing ratio and the film thickness increased fairly linearly with the number of cycles when using a 1:4 pulsing ratio. After annealing the films consisted of the hexagonal phase of $LiCoO_2$ according to both Raman and GIXRD (grazing incidence X-ray diffraction) measurements. Electrochemical characterization revealed that a 12% capacity loss was evident between charge and discharge cycles. For a film deposited with a 1:4 pulsing ratio, the capacity was only about 60% of the theoretical value. For a 1:2 ratio film, the capacity was even lower, which might be explained by the higher impurity contents in this film. With the 1:4 pulsing ratio, the composition of the films was $Li_{1.2}CoO_{3.5}$, as determined by elastic backscattering [88].

The potential cathode material lithium iron phosphate, $LiFePO_4$, has also been the subject of ALD studies [89,93] The material has been deposited at 300 °C on silicon substrates using ferrocene and ozone as precursors for the Fe_2O_3 subcycle, trimethylphosphate (TMPO) and water for PO_x and lithium *tert*-butoxide and water for the Li_2O/LiOH subcycle [89] Iron oxide and the phosphate were pulsed sequentially for five cycles, after which one cycle of Li_2O/LiOH was applied. The resulting films were amorphous and showed a linear increase in thickness as a function of deposited supercycles. The material could also be deposited onto carbon nanotubes (CNTs) [87] The CNT-based films were amorphous but crystallization to orthorhombic $LiFePO_4$ was observed after annealing in argon at 700 °C for 5 h. The Fe:P ratio in the annealed film was 0.9, as determined by EDX (energy-dispersive X-ray spectroscopy). Unfortunately, no compositional information on Li content was given. The $LiFePO_4$ film deposited onto CNTs showed good electrical performance, with clear redox peaks in a cyclic voltammetry curve at 3.5 V and 3.3 V (vs. Li/Li$^+$) and a discharge capacity of 150 mAh/g at 0.1 C [89]. Encouragingly, the material could maintain a discharge capacity of 120 mAh/g at 1 C even after 2000 cycles.

$LiFePO_4$ has also been deposited using metal-thd complexes [93]. Pulsing Lithd (lithium 2,2,6,6-tetramethyl-3,5-heptanedionate) and ozone between subcycles of Fe(thd)$_3$ + O_3 and TMPO + O_3 + H_2O resulted in stoichiometric $LiFePO_4$ when the fraction of Li_2CO_3 subcycles was 37.5%. The as-deposited films were amorphous but could be crystallized in 10/90 H_2/Ar atmosphere at 500 °C. These films were reported to show poor electrical conductivity, as expected with this material [19], however very little additional information was given. It should be noted that the same research group has also published an ALD process for the de-lithiated cathode material $FePO_4$ [94]. They reported an initial electrochemical capacity of 159 mAh/g for the as-deposited, amorphous 46 nm thick $FePO_4$ film. The capacity increased to 175 mAh/g after 230 charge-discharge cycles and after 600 cycles the capacity was still 165 mAh/g.

Lithium manganese spinel $Li_xMn_2O_4$ is an interesting cathode material for lithium-ion batteries due to its low volume change during (de)lithiation, high voltage and environmentally benign elements. The material has been deposited by ALD by Miikkulainen et al. using various methods [90]. Firstly, Mn(thd)$_3$ and ozone were used as precursors for manganese oxide and this process was combined with the Lithd + O_3 process for lithium incorporation. Interestingly, even with exceedingly small numbers of Li_2CO_3 subcycles, high Li$^+$ incorporation was achieved, with only a 5% Li_2CO_3 pulsing leading to a Li:Mn ratio of 1:1. This was in fact the maximum content of lithium obtained: using larger numbers of Li_2CO_3 subcycles led to a decrease in uniformity. To achieve the stoichiometric lithium level for $LiMn_2O_4$, 1% of Li_2CO_3 pulsing was sufficient. All the films showed the crystalline spinel phase as-deposited, with MnO_2 impurities present in the films with the lowest lithium concentrations. Crystalline spinel $LiMn_2O_4$ was also obtained by using LiO^{t}Bu and water as precursors, however little else was reported on this process.

The lithium manganese spinel process is unique in that while the growth rate of the films stays rather constant at below 0.3 Å/cycle, the lithium content increases very rapidly and reaches a high value with very small Li-subcycle numbers [90]. This indicates that the mechanism of this process differs significantly from conventional ternary ALD processes. Another clue about the mechanism was given by ToF-ERDA (time-of-flight elastic recoil detection analysis) elemental depth profiles which showed uniform film composition, albeit with a lithium deficiency on the film surface. To achieve such high lithium concentrations, either more than one monolayer should be deposited in one subcycle, or the growth should include a bulk component in addition to the normal surface reactions. Multilayer growth could be assumed to lead to lithium excess on the film surface, since lithium carbonate was always the last material deposited. Therefore, the bulk must be playing a role in the deposition process. Miikkulainen et al. postulated that the reduction needed for manganese to change from +IV in MnO_2 to +III/+IV in $LiMn_2O_4$ takes place during the Lithd pulse, which affects not only the surface but also deeper parts of the film [90]. The following ozone pulse removes organic residues from the surface and re-oxidizes the topmost manganese ions on the surface. This reaction then drives lithium ions deeper into the film, resulting in a uniform elemental distribution with a slightly lithium deficient surface.

Miikkulainen et al. continued their studies on $LiMn_2O_4$, using both LiO^tBu and water and Lithd and ozone exposures on MnO_2 at 225 °C [90]. Interestingly, 110 nm of manganese oxide could be converted to the spinel phase with only 100 cycles of the lithium carbonate process applied on top of the film. The carbonate was not present in the X-ray diffractogram. Lithiation was achieved to some extent also without ozone pulses. The manganese oxide films lithiated with LiO^tBu and water in a similar manner showed the best electrochemical storage properties, with a capacity of 230 mAh/g at 50 µA. The capacity retention was very good up to 550 cycles at 200 µA. Similar to $LiMn_2O_4$, vanadium oxide V_2O_5 could also be lithiated by pulsing either LiO^tBu and water or Lithd and ozone on top of the oxide film [90]. Using the Lithd and ozone precursors, lithium contents as high as 15 at% were obtained with only 100 cycles of the Li_2CO_3 process applied on the 200 nm oxide film.

Lithium sulphide, Li_2S, is an attractive cathode material for high capacity lithium-sulphur batteries. It has recently been deposited by ALD [91] and requires inert atmosphere during sample handling to prevent reactions with ambient air, in a similar way to pure Li_2O [95]. Li_2S has been deposited using LiO^tBu and hydrogen sulphide between 150 and 300 °C. Unlike most lithium containing processes, this one produced a constant growth rate over the whole deposition temperature range studied. The refractive index of the films was much lower than the value for bulk crystalline Li_2S, indicating a lower density of the films. The films were amorphous and could not be crystallized with annealing in inert atmosphere. Both XRF (X-ray fluorescence) and XPS (X-ray photoelectron spectroscopy) gave a Li:S ratio of 2:1, with no carbon contamination in the Li_2S layer. Thus, the reaction between the precursors was very efficient. The Li_2S films produced high capacities of 800 mAh/g when deposited onto mesocarbon microbeans and a somewhat lower capacity of 500 mAh/g when deposited directly onto a 2D Cu current collector. In both cases the Coulombic efficiency was ~100%, indicating that the material could indeed be used as a cathode in lithium-sulphur batteries. However, film thickness had a large effect on the capacity, with thicker films producing smaller capacities per gram, as Li_2S is rather insulating. In addition, reactions with the copper current collector affected the charge-discharge profiles, indicating the formation of Cu_xS.

4.2.2. Anodes

Table 5 presents examples of lithium-ion battery anode materials deposited by ALD. The selection of materials is quite a bit more limited than in the cathode case, most likely illustrating the consensus that improvements in battery energy density are more easily obtained with improved cathode materials. ALD-made TiO_2 has been studied as an anode material mostly in various 3D-constructions, illustrating the conformal coating ability of ALD [78,96–98]. Using 3D-structures the areal capacity of the titania anode can be greatly improved [96,97] Generally, TiO_2 can intercalate 0.5 Li, resulting in a capacity of 170 mAh/g [97] However, using nanomaterials more lithium can be intercalated and capacities of 330 mAh/g have been obtained with anatase nanotubes with a wall thickness of 5 nm [98]. These nanotubes also showed excellent capacity retention.

Table 5. Examples of conventional lithium-ion battery anode materials deposited by ALD. Abbreviations used: O^iPr = iso-propoxide, O^tBu = tert-butoxide, thd = 2,2,6,6-tetramethyl-3,5-heptanedionato, TPA = terephthalic acid, LiTP = lithium terephthalate.

Material	Precursors	T_{Dep} (°C)	Growth Rate (Å)/Binary Cycle	Capacity (mAh/g)	Ref.
TiO_2	$Ti(O^iPr)_4 + H_2O$	160	0.33	330	[98]
Li_xTi_yO	$LiO^tBu + Ti(O^iPr)_4 + H_2O$	225, 250	0.7 with Li:Ti pulsing ratio 1:1	40	[99–101]
LiTP	Lithd + TPA	200–280	3.0 at 200 °C, decreases with T_{Dep}	350	[102]

Attempts on the deposition of lithium titanate spinel, $Li_4Ti_5O_{12}$, have been made using both titanium tetrachloride $TiCl_4$ [99] and titanium tetra-iso-propoxide $Ti(O^iPr)_4$ as precursors [99–101]. In both cases, LiO^tBu was used as the lithium source and water was used as the oxygen source. Titanate films deposited using TiCl$_4$ reacted rapidly in air [99]. The films were amorphous as determined with X-ray diffraction and showed only very small amounts of lithium in ERDA measurements. In contrast, when using Ti(O^{i}Pr)$_4$ as the titanium source and applying a long pulse time for this precursor, uniform titanate films with higher lithium contents could be deposited [99]. These films also reacted with air, however the reaction was much slower than when using TiCl$_4$ as the titanium precursor. The growth rate of the films did not depend much on the pulsing ratio of the two metal precursors, being approximately 0.7 Å/cycle at 225 °C [99]. In another report using the same precursors, the growth rate was said to be slightly different at 0.6 Å/cycle at 250 °C [100]. For the process at 225 °C, ERDA measurements revealed that the lithium content of the films could be routinely tuned over a wide range by changing the metal pulsing ratio [99]. For example, with 33% lithium cycles the film stoichiometry was Li$_{1.19}$TiO$_{2.48}$ and the carbon and hydrogen impurity contents were low. XPS and ERDA revealed that in this material lithium was enriched on the film surface, most likely forming a carbonate layer: carbonate peaks were visible both in FTIR (Fourier transform infrared spectroscopy) and XPS [99,101]. Despite the carbonate formation, the films showed the Li$_4$Ti$_5$O$_{12}$ spinel phase in XRD measurements also in the as-deposited state. The crystallinity could be improved by annealing in nitrogen at 640–700 °C. The annealed titanate films showed electrochemical activity but the capacity remained low at 40 mAh/g [101]. However, this low value might be related to uncertainties in the calculation of film mass. For the film deposited at 250 °C, the Li:Ti ratio was reported as 2:1 with 44% lithium pulsing [100], which could indicate Li$_2$TiO$_3$ formation—a well-known impurity phase for Li$_4$Ti$_5$O$_{12}$ [103,104]. After annealing in argon at 850 and 950 °C these films showed XRD peaks belonging to Li$_4$Ti$_5$O$_{12}$ [100].

In addition to purely inorganic materials, ALD can also be used to deposit hybrid materials using organic molecules as the second precursor [105]. Lithium terephthalate (LiTP) has been deposited using Lithd and terephthalic acid as precursors between 200 and 280 °C [102]. This material has been proposed as a possible Li-ion battery anode due to its high theoretical capacity of 300 mAh/g and a low potential of 0.8 V (vs. Li$^+$/Li) [106]. The ALD process for LiTP showed saturation but no ALD window or constant growth rate as a function of the number of cycles [102]. The changing growth rate, accompanied by changes in the film density, could be related to the island growth mechanism of the film. The films were crystalline as deposited, which is unusual for ALD hybrid films. The films were electrochemically active and showed high rate capabilities with good capacity retention. The electrochemical properties of the films could further be enhanced by a protective LiPON layer on top of the electrode. The higher than theoretical capacity of 350 mAh/g is partly explained by difficulties in determining the electrode film thickness in electrochemical analyses. Recently, this anode material was combined with an organic cathode material dilithium-1,4-benzenediolate to produce a functioning atomic layer/molecular layer deposited thin film battery [107].

4.2.3. Solid Electrolytes

ALD has showcased its versatility in the deposition of potential solid electrolytes for lithium-ion batteries (Table 6). Many materials, both crystalline and amorphous, have been deposited and ionic conductivities of the order of 10^{-7} S/cm have been obtained with various materials. Of the traditional solid electrolytes, LiPON has been deposited using both ALD and PEALD [108,109].

Table 6. Examples of potential solid lithium-ion battery electrolyte materials deposited by ALD. Abbreviations used: LiHMDS = lithium hexamethyldisilazide, LiTMSO = lithium trimethyl silanolate, O^{t}Bu = *tert*-butoxide, TEOS = tetraethyl orthosilicate, TMPO = trimethyl phosphate, DEPA = diethyl phosphoramidate, OEt = ethoxide, thd = 2,2,6,6-tetramethyl-3,5-heptanedionato, TMA = trimethyl aluminum, La(FMAD)$_3$ = lanthanum tris(N,N-di-*iso*-propylformamidinate), TDMAZ = tetrakis(dimethylamido)zirconium, TDMA-Al = tris(dimethylamido) aluminum, HF-py = mixture of HF and pyridine.

Material	Precursors	T_{Dep} (°C)	Growth Rate (Å)/Binary Cycle	Ionic Conductivity (S/cm)	Ref.
Li$_x$Si$_y$O	LiHMDS + O$_3$	150–400	Varies with temperature	Not measured	[110,111]
	LiTMSO + O$_3$ + H$_2$O	175–300	1.5 at 200–300 °C	Not measured	[112]
	LiO^{t}Bu + TEOS + H$_2$O	225–300	0.5 at 250 °C	10^{-10}–10^{-9} at 30 °C	[113]
Li$_3$PO$_4$	LiO^{t}Bu + TMPO	225–300 [114] 250–350 [115]	0.7 at 225–275 °C [114], 0.69 at 300 °C [115]	10^{-8}–10^{-7} at RT [115,116]	[114–116]
LiPON	LiHMDS + DEPA	250–350	0.7 at 270–310 °C	6.6 × 10^{-7} at RT (9.7 at.% nitrogen)	[108]
	LiO^{t}Bu + H$_2$O + TMPO + N$_2$ plasma	250	1.05	1.45 × 10^{-7} (5.5 at.% nitrogen, increases with nitrogen contents)	[109]
Li$_x$Ta$_y$O	LiO^{t}Bu + Ta(OEt)$_5$ + H$_2$O	225	0.7 with Li:Ta pulsing ratio 1:6.	1.2 × 10^{-8} at RT	[117]
Li$_x$Nb$_y$O	LiHMDS + Nb(OEt)$_5$ + H$_2$O	235	~0.64 with Li:Nb pulsing ratio 1:2	Not measured	[118]
	LiO^{t}Bu + Nb(OEt)$_5$ + H$_2$O	235	~0.68 with Li:Nb pulsing ratio 1:2	6.4 × 10^{-8} at 30 °C	[119]
Li$_x$La$_y$Ti$_z$O	LiO^{t}Bu + La(thd)$_3$ + TiCl$_4$ + O$_3$ + H$_2$O	225	Varies with pulsing ratio	Not measured.	[120]
Li$_x$Al$_y$Si$_z$O	LiO^{t}Bu + TMA + TEOS + H$_2$O	290	1.0 with Li:Al pulsing ratio 6:10	10^{-9}–10^{-7} at RT, depends on Li contents	[121]
Li$_x$La$_y$Zr$_z$O:Al	LiO^{t}Bu + La(FMAD)$_3$ + TDMAZ + TMA + O$_3$	225	1.0 with Li:La:Zr:Al ratio 8:28:12:1	1 × 10^{-8} at RT for amorphous film	[122]
Li$_x$Al$_y$S	LiO^{t}Bu + TDMA-Al + H$_2$S	150	0.5 with Li:Al ratio 1:1	2.5 × 10^{-7} at RT	[123]
Li$_x$Al$_y$F	LiHMDS + TMA + HF-py	150	0.45 with Li:Al pulsing ratio 1:1	7.5 × 10^{-6} at RT	[51]
	LiO^{t}Bu + AlCl$_3$ + TiF$_4$	250	1 with Li:Al pulsing ratio 1:1	(3.5 ± 0.5) × 10^{-8} at RT	[124]

Lithium silicates can have reasonably high lithium-ion conductivities, especially in the amorphous state [125–127]. The silylamide precursor LiHMDS (lithium hexamethyldisilazide) provides a convenient route to lithium silicate deposition when combined with ozone [110,111]. The process exhibited good ALD behaviour at 250 °C, with saturation of both precursors seen and the film thickness increased linearly with the number of cycles. However, no ALD window was present and instead the growth rate increased from approximately 0.3 Å/cycle at 150 °C to 1.7 Å/cycle at 400 °C. This increase was explained with subsequent reaction mechanism studies [111]. The HMDS-ligand of the metal precursor reacts with surface hydroxyl groups, decomposing to different side products. Some of these side products are unreactive in the process but can still block active sites from the desired –Si(CH$_3$)$_3$ groups. At higher temperatures, the decomposition of the ligand to –Si(CH$_3$)$_3$ is enhanced and in addition desorption of unreactive products is faster. The deposited films were amorphous below 400 °C and showed only small amounts of carbon and hydrogen impurities, as determined by ERDA [110]. Notably, no nitrogen was detected in the film despite the lithium precursor being a silylamide. The Li:Si and Si:O ratios changed with deposition temperature but at 250 °C the film composition was Li$_2$SiO$_{2.9}$ which is very close to the lithium metasilicate Li$_2$SiO$_3$. The ionic conductivity of these films was not measured.

Recently, lithium silicates have been deposited also using LiTMSO (lithium trimethyl silanolate) [112] and lithium *tert*-butoxide [113]. With LiTMSO, to obtain good quality films both an ozone and a water pulse were needed after the metal precursor pulse [112]. It was postulated that the water generates hydroxyl groups on the surface of the silicate film, which are beneficial for the adsorption of LiTMSO. All the films were amorphous and the growth rate remained constant at 1.5 Å/cycle between 200 and 300 °C. Films deposited at low temperatures had significant hydrogen contents of 14 at.%. The amounts of both lithium and silicon increased with increasing deposition temperatures while the levels of impurities decreased but the Li:Si ratio remained at 2:1. No ionic conductivity information is available for these films.

Lithium *tert*-butoxide could deposit lithium silicates in combination with TEOS (tetraethyl orthosilicate) and water [113]. For these films, a Li:Si ratio close to Li_4SiO_4 was obtained at all deposition temperatures. The ionic conductivity of these films was quite low, reaching a maximum of 5×10^{-9} S/cm in films deposited at 250 °C.

LiPON, currently the most often used solid lithium-ion electrolyte material, was undoubtedly the stimulus for the ALD studies on lithium phosphate films. Li_3PO_4 can be deposited using either LiO^{t}Bu or LiHMDS as the lithium source and TMPO (trimethyl phosphate, Figure 9) as the phosphate precursor [114,115]. The LiO^{t}Bu + TMPO process showed a constant growth rate of approximately 0.7 Å/cycle between 225 and 275 °C [114]. However, no complete saturation was observed. The films were slightly crystalline and showed decreasing impurity levels at higher deposition temperatures in ERDA measurements. At 300 °C, the film composition was $Li_{2.6}PO_{3.7}$. The process utilizing LiHMDS is less than ideal, as the film growth rate varies strongly with deposition temperature, being 0.4 Å/cycle at 275 °C and 1.3 Å/cycle at 350 °C. At 300 °C, these films were close to stoichiometric lithium phosphate, being $Li_{2.8}PO_{3.9}$ as determined by ERDA. However, using LiHMDS as the lithium precursor led to higher carbon and hydrogen impurities than the LiO^{t}Bu process. Regardless of the lithium precursor, the phosphate films crystallized into the orthorhombic Li_3PO_4 phase during HTXRD measurements.

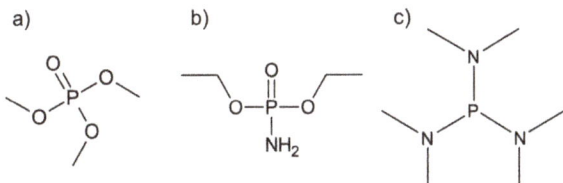

Figure 9. The structures of three phosphorus precursors used to deposit lithium phosphate and LiPON. (**a**) TMPO, or trimethyl phosphate; (**b**) DEPA, or diethyl phosphoramidate; (**c**) TDMAP, or *tris*(dimethylamino)phosphine.

Wang et al. [115] and Létiche et al. [116] have studied the Li_3PO_4 process using LiO^{t}Bu and TMPO in an effort to measure the lithium-ion conductivity of these films. Wang et al. reported an increasing growth rate as a function of the deposition temperature at 250–325 °C, which might be a result of the somewhat unsaturative behaviour of the process [114]. Electrochemical impedance spectroscopy showed that the films had rather good conductivities when deposited at 300 °C: 3.3×10^{-8} S/cm was extrapolated for a film with a composition of $Li_{2.8}PO_z$ (as determined by XPS) [115]. Similarly, Létiche et al. reported conductivities as high as 4.3×10^{-7} S/cm for Li_3PO_4 deposited at 300 °C [116]. These results are rather surprising, as it is common knowledge that lithium phosphate is generally no match for its nitrogen-doped counterpart LiPON and its conductivities of 10^{-8}–10^{-6} S/cm [32]. It appears that small film thicknesses can play a role in these high ionic conductivities [116]. Li_3PO_4 layers have been studied in contact with electrode materials [128,129] and it has been found that although the phosphate layer can decrease the electrode capacity, capacity retention is improved due to decreased transition metal dissolution and more stable SEI formation [128].

Recently, the deposition of LiPON was achieved both by thermal and plasma-enhanced ALD [108,109,130]. In the PEALD process, LiO^{t}Bu was used as the lithium source combined with a pulsing sequence of water, TMPO and nitrogen plasma [109]. Deposition of Li$_2$O/LiOH before exposure to TMPO resulted in less carbon impurities as compared to the process used by Hämäläinen et al. for Li$_3$PO$_4$ [114]. By using nitrogen plasma after the TMPO pulse, nitrogen could be incorporated into the films, causing the amorphization of the crystalline Li$_3$PO$_4$. In the thermal ALD process, the problems of nitrogen incorporation and nitrogen-phosphorous bond formation were resolved by using diethyl phosphoramidate, DEPA, a phosphate precursor with an amine group (Figure 9) [108]. By using DEPA with LiO^{t}Bu, nitrogen contents as high as 9.7 at.% were achieved. However, the thermal ALD process led to high carbon impurities from 9.9 to 13.3 at.% compared to virtually none in the PEALD process [109]. Both processes deposited conformal coatings on demanding substrates as required from a potential solid electrolyte material. In addition, very good electrochemical properties were realized with ionic conductivities of 1.45×10^{-7} S/cm for the PEALD process (5 at.% nitrogen) and 6.6×10^{-7} S/cm for the thermal ALD process (9.7 at.% nitrogen) [108,109]. The plasma-deposited LiPON has already been studied as a protecting layer for a conversion lithium-ion battery electrode. It was found that the LiPON layer enhanced the capacity retention of the electrode by providing both a high lithium-ion conductivity and mechanical support during cycling [131].

The most recent addition to the ALD LiPON processes was reported by Shibata, using TDMAP or *tris*(dimethylamino)phosphine (Figure 9) as the phosphorous source, LiO^{t}Bu as the lithium source and O$_2$ and NH$_3$ for oxidation and nitrification [130]. The high process temperatures of over 400 °C raise concerns of more CVD- than ALD-type film growth, especially combined with the changing growth rate as a function of cycles. However, no carbon impurities were found in the films with XPS. The N contents varied between 2 and 6 at.% and an ionic conductivity of 3.2×10^{-7} at 25 °C was obtained for these films.

Lithium tantalate, similarly to lithium niobate, is an interesting ferroelectric material [132,133]. Its amorphous form has also been suggested as a possible solid electrolyte material for lithium ions [117,134,135]. The material has been deposited by ALD using LiO^{t}Bu, Ta(OEt)$_5$ and water as precursors at 225 °C [117]. The film growth rate changed depending on the cycle ratio of the two binary processes, being 0.74 Å/binary cycle with a 1Li$_2$O + 6Ta$_2$O$_5$ pulsing sequence. Similarly, the lithium contents of the films changed drastically with pulsing ratio (Figure 10). Both XANES (X-ray absorption near edge structure) and XPS measurements revealed that the chemical environment of tantalum in the films was similar to that of tantalum in stoichiometric LiTaO$_3$. However, in the films deposited with the highest tantalum oxide pulsing ratios there were also some indications of a Ta$_2$O$_5$ phase. XPS also revealed some carbonate formation on the film surface. Less carbonate was formed on the surface of films deposited with high numbers of tantalum oxide subcycles, indicating that Ta$_2$O$_5$ was offering some protection for the lithium in the film against reactions with carbon dioxide in air. A lithium tantalate film with a composition of Li$_{5.1}$TaO$_x$ was studied with electrochemical impedance spectroscopy (EIS) [117]. The film showed a room temperature lithium-ion conductivity of 1.2×10^{-8} S/cm, which increased to 9.0×10^{-7} S/cm at 100 °C. The material has later been used as a protective layer on lithium nickel cobalt manganese oxide cathodes [135]. With 5 supercycles of LiTaO$_3$ (metal oxide pulsing ratio Li:Ta = 1:6), enhancements in both electrode capacity and cycling ability were obtained. Recently, LiTaO$_3$ films were also made using ALD and solid state reactions: Li$_2$CO$_3$ was deposited from Lithd and O$_3$ onto amorphous ALD-Ta$_2$O$_5$ and upon annealing at 750 °C in air crystalline LiTaO$_3$ was formed with low impurity levels and a Li:Ta ratio of 1.5:1 [104]. The ionic conductivity of this film has not been measured but it can be expected to remain small due to both the crystalline structure of the film and its close-to-stoichiometric content of lithium.

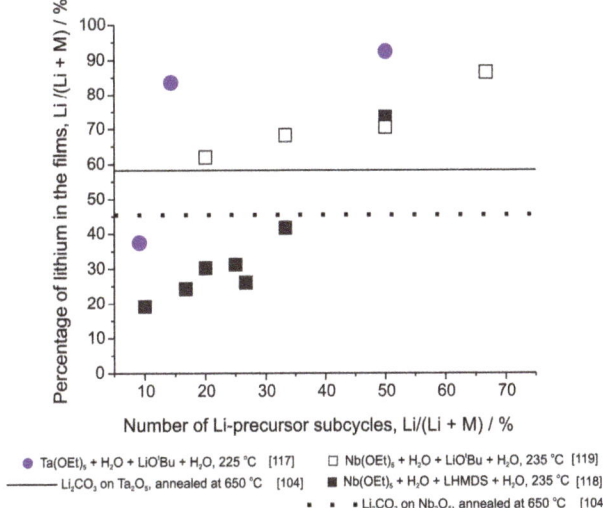

Figure 10. The amount of lithium cations deposited into lithium tantalate and lithium niobate films as a function of the amount of lithium containing subcycles. Data points obtained from [117–119]. The solid line represents the stoichiometry obtained for the lithium tantalate deposited by using 3000 cycles of Lithd and O_3 on 50 nm of Ta_2O_5 in Ref. [104]. The dashed line represents the stoichiometry obtained for the lithium niobate deposited by using same conditions but on Nb_2O_5 film in Ref. [104]. For stoichiometric $LiTaO_3$ and $LiNbO_3$, the lithium content is 50%.

Similar to lithium tantalate, lithium niobate thin films have also been deposited by ALD [118,119]. In the first paper, $LiNbO_3$ was deposited using $Nb(OEt)_5$, LiHMDS and water as precursors [118]. This work focused on the evolution of the lithium content in the films (Figure 10), on the epitaxial growth of the film on various surfaces and on the ferroelectric properties of the films. Later, the material was also deposited with LiO^tBu as the lithium source [119]. It is interesting to note that while the two processes use different lithium precursors, they produce films with the same stoichiometry when the pulsing sequence is Li:Nb = 1:1 (Figure 10). For other pulsing ratios, the LiO^tBu process seems to produce much higher lithium contents but this might be an artefact caused by the differing analysis methods used in these reports, XPS [119] and ToF-ERDA [118]. For the LiO^tBu process, the films with the lowest lithium contents showed the highest ionic conductivity of 6.4×10^{-8} S/cm at 30 °C [119]. In addition to these reports, $LiNbO_3$ has also been made with the same combination of ALD and solid state reactions as was mentioned for $LiTaO_3$ [104]. In this case also the post-deposition annealing of a bilayer of Li_2CO_3 and Nb_2O_5 produced a film with low impurity content. However, compared to the $LiTaO_3$ case, here the films were slightly lithium deficient (Figure 10).

The first truly quaternary lithium material deposited by ALD was lithium lanthanum titanate (LLT), reported by Aaltonen et al. in 2010 [120]. Thin films were deposited by combining binary ALD processes for TiO_2, La_2O_3 and $Li_2O/LiOH$ and were amorphous as deposited. In that work $TiCl_4$ was used as the titanium precursor and it was found that applying the $Li_2O/LiOH$ subcycle after the TiO_2 cycle resulted in rougher and less uniform films than when lithium was pulsed after the La_2O_3 subcycle. This is a clear indication that the pulsing order can have a large effect on the deposition of quaternary materials. As reactivity problems were also observed in the deposition of lithium titanate using $TiCl_4$ [99], the chloride precursor might be playing a large, thus far unknown role in these processes. These problems could be related to LiCl formation, for example. For the LLT deposition, a pulse sequence where 3 cycles of La_2O_3 were applied after one TiO_2 cycle and the number of lithium subcycles was varied, was used. The content of lithium in the film did not linearly follow the number of Li$_2$O/LiOH subcycles. This could mean that the reactivity of LiO^{t}Bu is lower on a Li$_2$O/LiOH surface compared to its reactivity on a La$_2$O$_3$ surface, a somewhat similar conclusion as was made in the deposition experiments on lithium aluminate [136]. Nevertheless, saturation as a function of the LiO^{t}Bu pulse length was observed [120]. The maximum lithium content reached with this pulsing scheme was approximately 20 at.%. The lanthanum content stayed constant in all experiments but the content of titanium decreased as a function of the increased lithium content. Under saturative conditions the film composition, as determined by ToF-ERDA, was Li$_{0.32}$La$_{0.30}$TiO$_z$. Interestingly, SIMS (secondary ion mass spectrometry) depth profiling seemed to indicate that lithium was somewhat concentrated onto the film-substrate interface, whereas in many cases lithium has been reported to preferably reside on the outer surface of the film [99,118]. However, this observation could be an artefact caused by sputtering during SIMS. The films could be crystallized by annealing in oxygen. The XRD diffractograms matched well with the reported peak positions of Li$_{0.33}$La$_{0.557}$TiO$_3$, however four peaks could not be identified [120].

Li$_x$Al$_y$Si$_z$O, another amorphous solid electrolyte, has been studied by Perng et al. [121]. The material belongs to the lithium aluminosilicate family, which includes materials with high lithium-ion conductivities with various metal ratios [30]. Lithium aluminosilicate was deposited by ALD using a pulsing sequence of Al$_2$O$_3$ from TMA and water, Li$_2$O/LiOH from LiO^{t}Bu and water and SiO$_2$ from TEOS and water [121]. With a pulsing sequence of Al:Li:Si = 10:6:4 the film thickness increased linearly with the number of supercycles. The lithium contents of the films, as determined by synchrotron ultraviolet photoemission spectroscopy (UPS), increased with increasing lithium oxide pulsing ratio but showed quite a lot of scattering (Figure 11). The deposited films were shown to be pinhole free and had ionic conductivities between 10^{-9} and 10^{-7} S/cm at room temperature, depending on the lithium content. Higher lithium contents led to higher conductivities but also increased the activation energy. It should be noted that the film thicknesses used in these experiments were very small, 10 nm and below. Larger thicknesses led to a lower ionic conductivity.

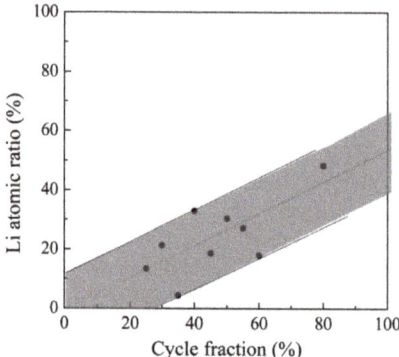

Figure 11. The lithium content in lithium aluminosilicate films as a function of the Li$_2$O cycle fraction $a/(a + b + c)$, as determined from syncrotron UPS spectra. Adapted with permission from [121]. Copyright 2014 The Royal Society of Chemistry.

Kazyak et al. have taken on the impressive task of depositing the garnet oxide Li$_7$La$_3$Zr$_2$O$_{12}$ by ALD [122]. This crystalline material is known to have a lithium-ion conductivity close to 10^{-3} S/cm at room temperature [137]. In order to stabilize the desired cubic phase at room temperature, the material was doped with alumina [122]. This resulted in an ALD process combining 8 subcycles of Li$_2$O, 28 subcycles of La$_2$O$_3$, 12 subcycles of ZrO$_2$ and 1 Al$_2$O$_3$ subcycle at 225 °C to obtain an amorphous film with metal ratios Li:La:Zr:Al = 52:27:19:2 (ideal composition 54:26:17:2). Interestingly, despite using ozone as the oxygen source for all subcycles, no Li$_2$CO$_3$ or La$_2$(CO$_3$)$_3$ formation was evident from XPS results. The thickness of the films increased linearly with the number of supercycles and good conformality was also obtained. The ionic conductivity of the as-deposited, amorphous film was 1.2×10^{-6} S/cm at 100 °C and did not differ between in-plane and through-plane measurements. By extrapolation, the conductivity was 10^{-8} S/cm at RT. The films could be crystallized to the cubic $Li_7La_3Zr_2O_{12}$ phase with annealing at 555 °C in inert atmosphere. A lithium-excess in the film and an extra lithium source were needed during the annealing due to lithium loss from the film. The annealed films had an island morphology, which prevented reliable conductivity measurements.

Although the majority of published ternary lithium ALD processes are for oxide materials, sulphides and fluorides have been studied as well [51,123]. Lithium aluminum sulphide Li_xAl_yS has been deposited using subcycles of Li_2S [91] (LiO^{t}Bu + H_2S) and Al_2S_3 [138] (tris(dimethylamido)aluminum(III) + H_2S). Using a 1:1 subcycle ratio resulted in a Li:Al ratio of 2.9:1 in films deposited at 150 °C, as determined with ICP-MS (inductively coupled plasma mass spectrometry) [123]. Evaluation of the metal ratio from QCM (quartz crystal microbalance) data, assuming stoichiometric growth, resulted in a metal ratio of 3.5:1 which is reasonably close to the value from ICP-MS. The ternary sulphide growth was linear as a function of cycles, with a reported growth rate of 0.50 Å/cycle. The growth rate during the Li_2S subcycle seemed somewhat lower in the ternary process than the one reported for the binary process [91]. This difference was not commented on in ref. [123] but it most likely originates from different starting surfaces. A 50 nm Li_xAl_yS film was measured to have a room temperature ionic conductivity of 2.5×10^{-7} S/cm, which is among the best quoted conductivities of ALD-made films [108,109,117,121,123]. The sulphide was studied as an artificial SEI-layer on metallic Li and it was found to effectively stabilize the interface between the metal anode and an organic liquid electrolyte [123]. In addition, the coating decreased lithium metal dendrite formation during cycling, which considerably improves the safety of lithium metal anodes.

Multicomponent, lithium-containing fluorides have not been studied extensively by ALD [51,124,139]. Still, Li_3AlF_6, has been deposited using multiple processes. This material will be discussed in the next section on fluoride deposition using ALD.

5. Atomic Layer Deposition of Metal Fluorides

Metal fluorides have been of interest to ALD chemists since the beginning of the 1990s. In the very beginning, doping of electroluminescent materials with fluorine was studied [140] and soon after the first report on depositing CaF_2, ZnF_2 and SrF_2 was published [141]. For the first two decades, metal fluorides were studied mainly because of their optical properties, namely low refractive indices and low absorption in the UV range [142]. However, with the rise of lithium-ion battery related ALD research, the potential of ALD metal fluorides in batteries has also been recognized [51,143–145]. Still, very few results on using atomic layer deposited metal fluoride thin films in lithium-ion batteries is available at this time.

Table 7 summarizes all reported ALD fluoride processes. Electrochemical analysis results are reported when available. The processes are divided into sections based on the fluorine precursor used. The materials are listed in the order of main groups followed by transition metals and lanthanides in the order of atomic numbers. For discussion purposes, a somewhat historical approach has been taken in the following subchapters. Besides this review, ALD of metal fluorides has been discussed in the academic dissertations of Pilvi [142], Lee [146] and Mäntymäki [79].

Table 7. ALD processes reported for fluoride materials. Abbreviations used: LiHMDS = lithium hexamethyldisilazide, HF-py = mixture of HF and pyridine, EtCp = ethylcyclopentadienyl, thd = 2,2,6,6-tetramethyl-3,5-heptanedionato, TMA = trimethyl aluminum, Ac = acetate, DEZ = diethylzinc, TEMAZ = tetrakis(ethylmethylamido) zirconium, O^{t}Bu = *tert*-butoxide, TDMAH = tetrakis(dimethylamido) hafnium, hfac = 1,1,1,5,5,5-hexafluoro-2,4-pentanedionato.

Material	Precursors	T_{Dep} (°C)	Growth Rate (Å)/Cycle	Ref.
Fluorides Using HF as the Fluorine Precursor				
LiF	LiHMDS + HF-py	150	0.5	[147]
MgF$_2$	Mg(EtCp)$_2$ + HF-py	150	0.4	[147]
MgF$_2$	Mg(EtCp)$_2$ + HF	100–250	Varies, 0.6 at 100 °C	[148]
CaF$_2$	Ca(thd)$_2$ + HF/NH$_4$F	300–400	0.2 at 320–400 °C	[141]
SrF$_2$	Sr(thd)$_2$ + HF/NH$_4$F	260–320	Varies, 0.6 at 300 °C	[141]
AlF$_3$	TMA + HF-py	75–300	Varies, 1.0 at 150 °C	[143]
AlF$_3$	TMA + HF	100–200	Varies, 1.2 at 100 °C	[149]
MnF$_2$	Mn(EtCp)$_2$ + HF-py	150	0.4	[147]
ZnF$_2$	Zn(Ac)$_2$·2H$_2$O + HF/NH$_4$F	260–320	0.7 at 260–300 °C	[141]
ZnF$_2$	DEZ + HF-py	150	0.7	[147]
ZrF$_4$	TEMAZ + HF-py	150	0.9	[147]
ZrF$_4$	Zr(O^{t}Bu)$_4$ + HF-py	150	0.6	[147]
HfF$_4$	TDMAH + HF-py	150	0.8	[147]
Li$_x$Al$_y$F	LiHMDS + TMA + HF-py	150	0.45 with Li:Al pulsing ratio 1:1	[51,146]
Fluorides Using Metal Fluorides as the Fluorine Precursor				
LiF	Lithd + TiF$_4$	250–350	1.0 at 325 °C	[150]
LiF	Mg(thd)$_2$ + Lithd + TiF$_4$	300–350	1.4 at 325 °C	[151]
LiF	LiO^{t}Bu + WF$_6$	150–300	–	[152]
LiF	LiO^{t}Bu + MoF$_6$	150–300	2.6	[152]
LiF	LiO^{t}Bu + TiF$_4$	200–300	0.5 at 250 °C	[124]
MgF$_2$	Mg(thd)$_2$ + TiF$_4$	250–400	Varies, 1.6 at 250 °C	[153]
MgF$_2$	Mg(thd)$_2$ + TaF$_5$	225–400	Varies, 1.1 at 225–250 °C	[154]
CaF$_2$	Ca(thd)$_2$ + TiF$_4$	300–450	Varies, 1.6 at 300–350 °C	[155]
AlF$_3$	AlCl$_3$ + TiF$_4$	160–340	Varies, 0.75 at 240 °C	[156]
AlF$_3$	TMA + TaF$_5$	125–350	Varies, 1.9 at 125 °C	[145]
YF$_3$	Y(thd)$_3$ + TiF$_4$	175–325	Varies, 1.3–1.5 at 200–300 °C	[157]
LaF$_3$	La(thd)$_3$ + TiF$_4$	225–350	Varies, 5.2 at 225–250 °C	[158]
Li$_x$Al$_y$F	LiF + Al(thd)$_3$ + TiF$_4$	250–350	–	[139]
Li$_x$Al$_y$F	LiO^{t}Bu + AlCl$_3$ + TiF$_4$	250	1	[124]
AlW$_x$F$_y$	TMA + WF$_6$	200	1–1.5	[144,159]
Fluorides Using Other Processes				
LiF	MgF$_2$ + Lithd	275–325	–	[151]
MgF$_2$	Mg(thd)$_2$ + Hhfac + O$_3$	–	0.38	[160]
CaF$_2$	Ca(hfac)$_2$ + O$_3$	300	0.3	[160]
CaF$_2$	Ca(thd)$_2$ + Hhfac + O$_3$	250–350	0.4	[160]
AlF$_3$	TMA + SF$_6$ plasma	50–300	Varies, 0.85 at 200 °C	[161]
LaF$_3$	La(thd)$_3$ + Hhfac + O$_3$	–	0.49	[160]
Li$_x$Al$_y$F	AlF$_3$ + Lithd	250–300	–	[139]

5.1. ALD of Metal Fluorides Using HF as the Fluorine Source

ZnF$_2$, SrF$_2$ and CaF$_2$ films, reported in the first paper on ALD of fluorides in 1994, were deposited using HF as the fluorine source [141]. The HF gas was generated in the reactor in situ by thermal decomposition of ammonium fluoride, NH$_4$F. Thus, there was no need to store and handle large amounts of gaseous HF. An added benefit of this method was that excess HF can be condensed again inside the reactor as ammonium fluoride, without the gas entering and damaging the vacuum pump. Metal thd-complexes were used as precursors for strontium and calcium and zinc fluoride was deposited using zinc acetate. All the films were close to stoichiometric and polycrystalline, with carbon impurities of the order of 0.5 at.%. For the calcium and strontium fluoride processes, the growth rates decreased with increasing deposition temperatures (Figure 12b).

The work on fluoride deposition using HF has been continued by many groups, including Hennessy et al., who have deposited magnesium and aluminum fluoride films using anhydrous HF with bis(ethylcyclopentadienyl)magnesium and TMA as metal precursors [148,149]. Magnesium fluoride is an interesting material due to its large band gap and low refractive index [142,153]. Aluminum fluoride is a material with a similar variety of possible optical applications [149,162–164]. In addition, as already mentioned, AlF_3 is a potential artificial SEI-layer for protecting both cathodes and anodes [47,143,145,165–167]. Thus, AlF_3 has become a much studied ALD material in the past few years [143,145,147,149,156]. Magnesium fluoride showed growth rates of 0.6 to 0.3 Å/cycle in the deposition temperature range of 100 to 250 °C (Figure 12a) [148]. AlF_3 showed a similar decrease in growth rate, being 1.2 Å/cycle at 100 °C and 0.5 Å/cycle at 200 °C (Figure 12c) [149]. MgF_2 films were crystalline and showed small amounts of carbon and oxygen impurities and a slight fluorine deficiency in XPS measurements (Figure 12d) [148]. AlF_3 films were amorphous, with 1–2 at.% of oxygen [149]. The aluminum fluoride films were stoichiometric based on XPS measurements. The anhydrous HF required an unconventionally long purging time to obtain good film uniformity. It was speculated that multilayer physisorption might be the cause of this effect. However, it has been reported in another publication that MgF_2 does not readily adsorb HF during the ALD growth process [147].

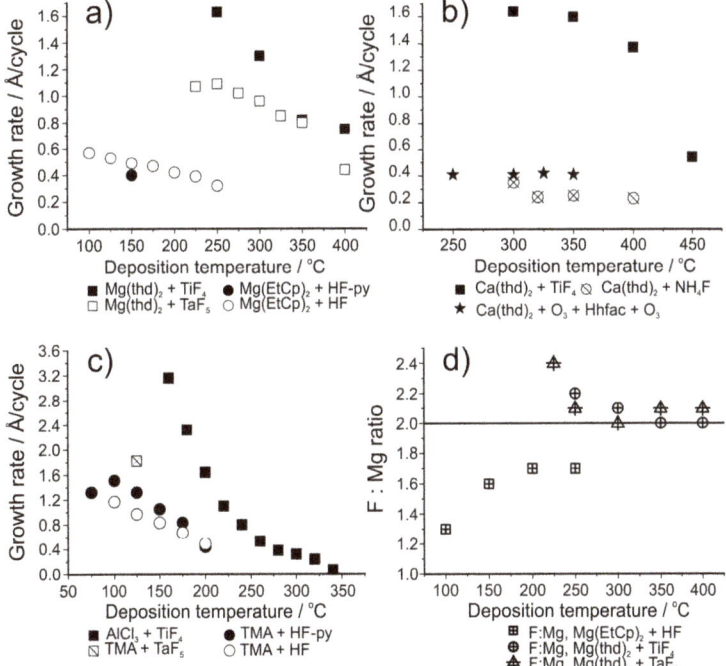

Figure 12. (a) Growth rates of MgF_2 films as a function of deposition temperature using different precursor combinations; (b) Growth rates of CaF_2 films as a function of deposition temperature using different precursor combinations; (c) Growth rates of AlF_3 films as a function of deposition temperature using different precursor combinations; (d) F-Mg ratios of MgF_2 films deposited at different temperatures and with different precursor combinations. The HF-process was analysed with XPS and the TiF_4 and TaF_5 processes with ERDA. Data obtained from References: (a) [147,148,153,154]; (b) [141,155,160]; (c) [143,145,149,156]; (d) [148,153,154].

A number of metal fluorides have recently been deposited by Lee et al. using HF, including AlF$_3$ [143], LiF, ZrF$_4$, ZnF$_2$ and MgF$_2$ [147]. Of these materials, lithium fluoride is of special interest because of its band gap of approximately 14 eV and low refractive index of 1.39 at 580 nm, much like AlF$_3$ as discussed previously [168,169]. For the deposition of these fluorides, HF was generated from a solution containing 30 wt.% of pyridine and 70 wt.% HF ("Olah's reagent") to mitigate the safety concerns of anhydrous, gaseous HF. This solution is in equilibrium with gaseous HF, with no pyridine detected in the gas phase and provides a safer alternative to anhydrous HF [143,147]. Metal precursors used included a diethylcyclopentadienyl complex for magnesium, a silylamide for lithium and an alkylamide for zirconium. All processes resulted in saturation at 150 °C, with growth rates below 1 Å/cycle (Figure 12). All films, except AlF$_3$ and ZnF$_2$, were crystalline. Generally, the films contained less than 2 at.% of oxygen impurities, as determined with XPS. Only ZrF$_4$ contained some carbon impurities in addition to the oxygen. The films appeared somewhat fluorine deficient, however this is speculated to be a result of preferential fluorine sputtering during the XPS measurement [143,147]. The AlF$_3$ deposition from TMA and HF showed an interesting etching reaction: above 250 °C the precursor pulses etched the AlF$_3$ film [143,149]. This reaction has later been exploited in developing atomic layer etching processes [170–172]. The AlF$_3$ process has been successfully utilized in protecting freestanding LiCoO$_2$/MWCNT (multi-walled carbon nanotube)/nanocellulose fibril electrodes, showing better protection at high potentials compared to the more common artificial SEI material Al$_2$O$_3$ [173].

Li$_3$AlF$_6$ has been deposited using subcycles of AlF$_3$ and LiF using TMA and HF-pyridine and LiHMDS and HF-pyridine as precursors [51]. One subcycle of AlF$_3$ and one subcycle of LiF were used at 150 °C and monoclinic Li$_3$AlF$_6$ was obtained with a growth rate of 0.9 Å/cycle [51,146]. The films had a Li:Al ratio of 2.7:1, as determined with ICP-MS and carbon, silicon and oxygen impurities were below the XPS detection limit. The films had an ionic conductivity of 7.5 × 10^{-6} S/cm at room temperature [146], which is similar to the first reports from the literature on thermally evaporated amorphous Li$_3$AlF$_6$ films [12,13]. Interestingly, changing the pulsing ratio to three subcycles of AlF$_3$ and one LiF resulted in the same metal ratio in the as-deposited film as the pulsing ratio of 1:1, suggesting a similar conversion reaction as we have seen in our experiments on LiF [151] and Li$_3$AlF$_6$ [139].

5.2. ALD of Metal Fluorides Using Metal Fluorides as the Fluorine Source

The deposition of many metal fluorides has been studied at the University of Helsinki using TiF$_4$ as the fluorine source, including materials such as MgF$_2$, LaF$_3$ and CaF$_2$ [153,155,157,158]. TiF$_4$ is a relatively safe alternative to HF, since it is a solid at room temperature. It possesses relatively high volatility and thermal stability, combined with high reactivity, which are vital attributes for an effective ALD precursor. The use of TiF$_4$ is made possible by a net ligand exchange reaction with a metal thd-complex (Equation (2)):

$$4M(thd)_x \, (g) + xTiF_4 \, (g) \rightarrow 4MF_x \, (s) + xTi(thd)_4 \, (g) \qquad (2)$$

Other volatile side products, such as TiF$_x$(thd)$_{4-x}$ can form in addition. Recently this precursor was also demonstrated in an ALD process used in conjunction with AlCl$_3$ [156].

Generally, metal fluorides deposited using TiF$_4$ as the fluorine source show a decrease in growth rate as the deposition temperature is increased (Figure 12) [155]. This decrease has been proposed to be due to a decrease in the TiF$_x$ adsorption density but this has not been verified experimentally. Using TiF$_4$ as the fluorine source leads to higher growth rates compared to the use of HF, possibly due to the formation of the fluoride in question during both precursor pulses [155,156]. When using TiF$_4$ as the fluorine source, the film growth usually shows saturation with respect to the fluorine precursor (Figure 13b) but the metal precursor can show either poor [150,153,157] or good saturation [158], depending on the material deposited (Figure 13a). Pilvi et al. postulated that the reason for the non-saturative behaviour might be either slow kinetics or an enhancement of metal precursor decomposition caused by the TiF$_x$-surface groups [153]. Films deposited with TiF$_4$ are generally very close to stoichiometric, as determined with ERDA (Figure 12d). Titanium is often found as an impurity in the deposited films but usually in only very small amounts [150,153,155]. Still, this impurity can limit the UV transmittance of these films when optical applications are the goal. The impurity level decreases as the deposition temperature is increased but at the same time film roughness increases resulting in more scattering of UV-light [153,157,158].

In an effort to obtain even purer metal fluoride films for optical applications, deposition of MgF$_2$ has been studied using TaF$_5$ as the fluorine source [154]. The growth process is very similar to that using TiF$_4$ (Figure 12a) although with using TaF$_5$ saturation with respect to the Mg(thd)$_2$ pulse length is observed. The films contained lower metal impurity levels than those deposited with TiF$_4$ and in addition the films were much smoother at high deposition temperatures [153,154]. This low roughness resulted in improved optical properties.

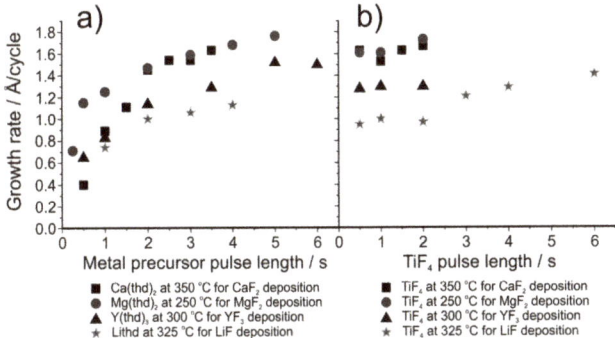

Figure 13. (a) Saturation curves for the metal precursor for CaF$_2$, MgF$_2$, YF$_3$ and LiF deposition processes; (b) Saturation curves for the fluorine precursor TiF$_4$ for CaF$_2$, MgF$_2$, YF$_3$ and LiF processes. Data obtained from [150,153,155,157].

Following the work of Pilvi et al., a process for the deposition of LiF was developed [150]. Lithd and TiF$_4$ were used as precursors and the deposition temperature was varied between 250 and 350 °C. Crystalline LiF films with only small proportions of impurities were obtained at all deposition temperatures. Saturation of the growth was not found for Lithd at 325 °C (Figure 13a). TiF$_4$, on the other hand, showed saturation type behaviour between 0.5 and 2.0 s pulses. With longer pulse times, the growth rate increased linearly. Such an increase has not been observed with other processes utilizing TiF$_4$, however in these studies the TiF$_4$ pulse lengths have been limited to 2 s or less (Figure 13b) [153,155,157,158].

The pulsing sequence Mg(thd)$_2$ + TiF$_4$ + Lithd + TiF$_4$ can produce LiF thin films between 300 and 350 °C [151]. Unlike the Lithd + TiF$_4$ process of ref. [150], this sequence shows both an ALD window between 325 and 350 °C and saturation with respect to both Lithd and TiF$_4$ (Figure 14). The growth rate at 325 °C was 1.4 Å/cycle, as opposed to 1.0 Å/cycle in the previous LiF process. All the films were again highly crystalline, with the film roughness being 19–20 nm for 70–80 nm films regardless of the deposition temperature. ToF-ERDA measurements showed the films to be very pure LiF, with only minute amounts of Mg and Ti impurities. C and H formed the largest part of impurities, however both were below 1 at.% in the deposition temperature range of 300–350 °C. It is surprising that despite the use of Mg(thd)$_2$ in the pulsing sequence, no magnesium ended up in the LiF films. We have proposed a mechanism to explain the deposition process (Equations (3)–(6)). In Equation (3), Mg(thd)$_2$ and TiF$_4$ deposit MgF$_2$ as has been previously reported [153]. In Equation (4), Li$^+$ from Lithd replaces Mg^{2+} in the fluoride film and forms LiF. Magnesium leaves the films as Mg(thd)$_2$, because the low levels of O,

C and H impurities imply that virtually no ligand decomposition is occurring during the growth. This type of fluoride-to-β-diketonate ligand exchange might at first seem unexpected, however it has been reported that metal oxides can be dry etched using β-diketone vapours to form volatile β-diketonato complexes of metal ions [174]. After the removal of magnesium, Lithd adsorbs onto the formed lithium fluoride (Equation (5)) and is converted to LiF during the last TiF$_4$ pulse (Equation (6)).

$$2Mg(thd)_2 \text{ (ads)} + TiF_4 \text{ (g)} \rightarrow 2MgF_2 \text{ (s)} + Ti(thd)_4 \text{ (g)} \quad (3)$$

$$MgF_2 \text{ (s)} + 2Lithd \text{ (g)} \rightarrow 2LiF \text{ (s)} + Mg(thd)_2 \text{ (g)} \quad (4)$$

$$LiF \text{ (s)} + Lithd \text{ (g)} \rightarrow Lithd \text{ (ads)} \quad (5)$$

$$4Lithd \text{ (ads)} + TiF_4 \text{ (g)} \rightarrow 4LiF \text{ (s)} + Ti(thd)_4 \text{ (g)} \quad (6)$$

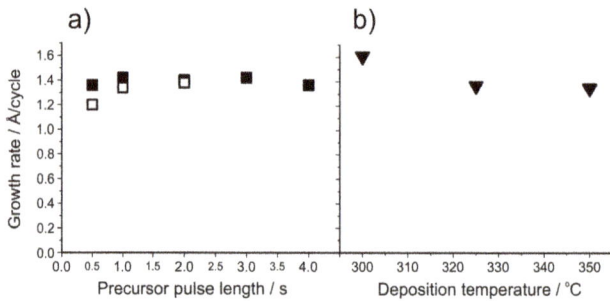

Figure 14. (a) Growth rate of LiF films as a function of Lithd (black squares) and TiF$_4$ (white squares) pulse lengths at 325 °C; (b) Growth rate of LiF films as a function of deposition temperature. Data from [151].

Xie et al. have studied the use of LiO^{t}Bu instead of Lithd with TiF$_4$ as the fluorine source [124]. This precursor combination led to the deposition of crystalline LiF between 200 and 300 °C. The maximum growth rate of 0.5 Å/cycle was achieved at 250 °C. The films had a refractive index close to 1.4 and a Li:F ratio of 1:0.97. Little else has been reported on the process thus far.

Similar to the methodology of using TiF$_4$ or TaF$_5$ as fluorine sources, Mane et al. have reported on the deposition of LiF using LiO^{t}Bu and either WF$_6$ or MoF$_6$ as the fluorine source [152]. Film growth took place between 150 and 300 °C, with amorphous films being deposited at the lowest temperature. This is an interesting finding, as using LiHMDS and HF-py at 150 °C led to crystalline LiF films [147]. With MoF$_6$ as the fluorine source the films had a growth rate of 2.6 Å/cycle, which is much higher than that obtained with the Lithd + TiF$_4$ process [150]. The films showed a 1:1 ratio of Li and F, with very small amounts of oxygen and carbon impurities. Most importantly, metal impurities were not detected with XPS.

After the success of alkaline and alkaline earth metal fluoride deposition, our group studied Al(thd)$_3$ and TiF$_4$ as precursors for AlF$_3$ [156]. However, this precursor combination led to no film growth on silicon and aluminum oxide. Our assumption is that this lack of reactivity has to do with the presence of aluminum-oxygen bonds in the Al(thd)$_3$ complex. Therefore, another aluminum precursor was needed in combination with TiF$_4$. AlCl$_3$ is a widely used ALD precursor, with no oxygen present in the molecule. In addition, TiCl$_4$ is a well-known, volatile ALD precursor [175,176], which is encouraging considering the expected ligand-exchange reaction taking place between AlCl$_3$ and TiF$_4$. Thus, this combination was studied for the deposition of AlF$_3$ [156]. Film growth was observed between 160 and 340 °C (Figure 12c). The saturation of the growth rate was studied at 160, 200 and 240 °C (Figure 15a). TiF$_4$ showed similar behaviour as in the LiF case [150], with an increased growth rate with the longest pulse times. AlCl$_3$, on the other hand, showed an opposite trend with growth rates decreasing as a function of pulse time. AlCl$_3$ vapour has been reported to enhance the volatility of AlF$_3$ [177], which might explain the etching-type behaviour seen in this process. As already mentioned, a similar decrease in the growth rate of AlF$_3$ was also observed in the TMA + HF-py process. Despite these effects, the AlF$_3$ growth rate remained constant with different cycle numbers at all temperatures studied (Figure 15b) [156]. ToF-ERDA measurements revealed that the films contained decreasing amounts of Cl and Ti impurities as the deposition temperature was increased, both being well below 1 at.% at 280 °C. However, the H and O impurities showed the opposite trend, with the films deposited at 280 °C containing up to 6 at% of oxygen.

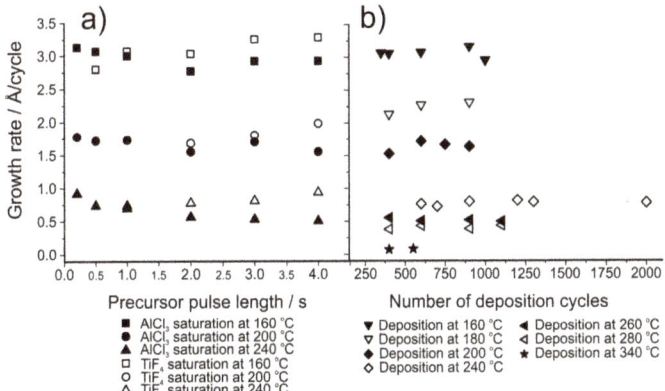

Figure 15. (a) Growth rate of AlF$_3$ films as a function of AlCl$_3$ (black) and TiF$_4$ (white) pulse lengths at 160, 200 and 240 °C; (b) Growth rate of AlF$_3$ films as a function of deposition cycles at various deposition temperatures. The AlCl$_3$ pulse time was 0.5 s and the TiF$_4$ pulse time was 1 s. Data from [156].

Jackson et al. have attempted to combine the methods discussed above for the deposition of AlF$_3$ by using TMA and TaF$_5$ as precursors [145]. This approach is somewhat questionable, as it has been reported that combining TMA with metal halides generally leads to metal carbide deposition [178–180]. In addition, a similar ligand exchange reaction between TMA and TaF$_5$, as was depicted in Equation (2), should produce pentamethyl tantalum which has been reported to be unstable [181,182]. Indeed, significant amounts of TaC$_x$ were deposited at elevated temperatures [145]. At 125 °C the content of tantalum impurity was decreased and the process showed ALD-like behaviour. The films contained approximately 20 at.% of oxygen, meaning that the process is unable to deposit good quality AlF$_3$. Despite the large amounts of impurities, the deposition of this material onto a high-voltage lithium-ion battery cathode nickel-manganese-cobalt oxide (NMC) led to significant improvements in its rate performance (Figure 16).

Similar to Jackson et al., Park et al. used TMA with WF$_6$ to deposit an amorphous composite fluoride composed of AlF$_3$ and metallic W and WC$_x$ [144,159]. The material was studied as an artificial SEI layer for LiCoO$_2$ cathodes and was found to improve the cycling properties of the material. It appeared that the composite nature increased the electron conductivity of the fluoride layer while still retaining its chemical inertness.

Our group has also studied ternary fluoride deposition using metal fluorides as precursors [139]. The deposition of Li$_3$AlF$_6$ proved complicated when using the binary processes of LiF and AlF$_3$ described in Refs. [150,151,156]: uncontrollable conversion of AlF$_3$ to LiF by Lithd was a threat to using the ALD subcycle approach. More importantly, exposing LiF to AlCl$_3$ could result in undesirable LiCl formation. Thus, attempts at depositing Li$_3$AlF$_6$ were made using two processes (Figure 17) [139]. In Process 1, Al(thd)$_3$ and TiF$_4$ were pulsed sequentially onto LiF thin films and a conversion reaction to the ternary Li$_3$AlF$_6$ took place during the deposition process. In Process 2, AlF$_3$ films were exposed to Lithd vapour in an ALD reactor (described in more detail in section "5.3. Other Approaches to ALD of Metal Fluorides"). In Process 1, Al(thd)$_3$ was used instead of AlCl$_3$ to avoid Li$^+$ contact with Cl$^-$ from AlCl$_3$. Despite the prior knowledge that Al(thd)$_3$ and TiF$_4$ do not produce AlF$_3$ on silicon substrates [156], a reaction did occur between these precursors when pulsed onto LiF films. The fluorides mixed together during the deposition process, resulting in crystalline Li$_3$AlF$_6$ with crystalline LiF residues. High deposition temperatures together with long Al(thd)$_3$ pulses resulted in less LiF impurity in the film, as observed with GIXRD. However, these same conditions worsened the visual appearance of the films. ToF-ERDA revealed that even in the best samples, the content of Al was very low, although Li$_3$AlF$_6$ was clearly visible in the X-ray diffractograms. In addition, the level of titanium impurity was high. Thus, it was concluded that in the end Process 1 was not efficient in depositing Li$_3$AlF$_6$.

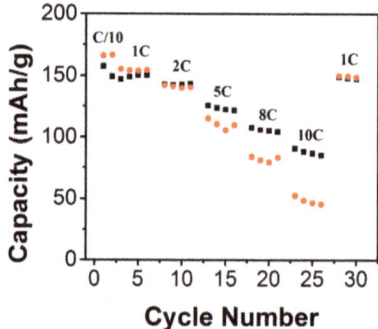

Figure 16. Gravimetric capacity of coin cells as a function of cycle number with different discharge rates. Black squares denote an NMC-cathode coated with TMA + TaF$_5$. Red circles denote uncoated NMC. Reprinted with permission from [145]. Copyright 2016 American Vacuum Society.

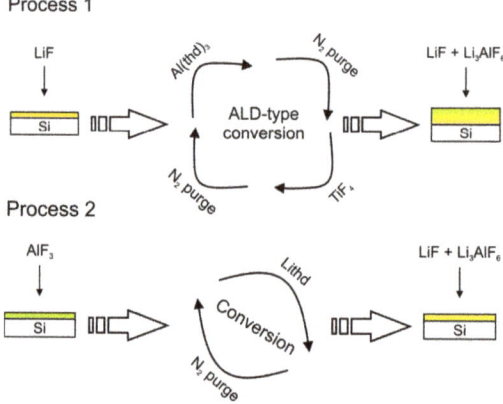

Figure 17. Process 1 utilizes Al(thd)$_3$ and TiF$_4$ in a conversion reaction to form Li$_3$AlF$_6$ out of LiF thin films. Process 2 uses a conversion reaction between ALD-made AlF$_3$ and the lithium precursor Lithd to deposit Li$_3$AlF$_6$. Reprinted with permission from [139]. Copyright 2017 Elsevier.

Recently, Xie et al. published their results on the deposition of LiAlF$_4$ [124]. By combining the processes for LiF and AlF$_3$ using LiO^{t}Bu, AlCl$_3$ and TiF$_4$ as precursors they were able to deposit amorphous LiAlF$_4$ with a Li:Al ratio of 1.2:1 and an ionic conductivity of 3.5×10^{-8} S/cm. The LiAlF$_4$ film was used as a protective layer on top of a lithium nickel manganese cobalt oxide cathode and was found to improve the stability of the cathode without sacrificing its rate performance.

5.3. Other Approaches to ALD of Metal Fluorides

Putkonen et al. provided other interesting pathways for avoiding the use of HF in fluoride deposition by depositing metal fluorides through oxide chemistry [160]. They found that by using the fluorinated β-diketonate precursor Ca(hfac)$_2$ and ozone as precursors CaF$_2$ is deposited, instead of CaCO$_3$ that is formed when the non-fluorinated β-diketonate Ca(thd)$_2$ is used together with ozone [183]. The fluoride films had a growth rate of 0.3 Å/cycle at 300 °C and were close to stoichiometric, although approximately 5 at.% of oxygen was present in the films [160]. An even more interesting approach to CaF$_2$ used the non-fluorinated metal precursor Ca(thd)$_2$ in combination with ozone and the Hhfac molecule. First, Ca(thd)$_2$ was pulsed onto the substrate followed by an ozone pulse, resulting in CaCO$_3$ deposition. Then Hhfac, which is known to adsorb to surfaces, was pulsed followed again by an ozone pulse. Ozone breaks down the Hhfac on the surface, providing fluoride ions which react with the calcium ions, resulting in a conversion reaction from CaCO$_3$ to CaF$_2$. This process provided a growth rate of 0.4 Å/cycle between 250 and 350 °C, which is close to the rate of CaCO$_3$ deposition from Ca(thd)$_2$ and ozone in similar conditions. The CaF$_2$ films were polycrystalline and close to stoichiometric and the amount of oxygen was below the detection limit of Rutherford backscattering spectroscopy (RBS). The same approach was reported to be successful also for MgF$_2$ and LaF$_3$ deposition.

Vos et al. recently reported on an interesting process for depositing AlF$_3$ using TMA and SF$_6$ plasma [161]. Plasma-enhanced ALD processes have not been previously available for fluoride deposition. However, F-containing plasmas, such as CF$_4$, have been widely used for etching processes. In this work, SF$_6$ was used because it is stable and non-toxic. Inductively coupled SF$_6$ plasmas contain species such as F, F$_2$, SF^{5+} and F$^-$ with S and S$^+$ as minor components. The AlF$_3$ deposition showed saturation with both TMA and the plasma with good film uniformity and conformality. As with all other AlF$_3$ processes, the film growth rate decreased with increasing deposition temperature, being 0.85 Å/cycle at 200 °C. All the films were amorphous with very low S and O impurity levels and an Al:F ratio close to stoichiometric.

After noting that the pulsing sequence Mg(thd)$_2$ + TiF$_4$ + Lithd + TiF$_4$ produced LiF with little to no Mg impurities, our group studied the conversion reaction of MgF$_2$ films upon exposure to Lithd [151]. As it turned out, with high enough Lithd doses MgF$_2$ films of 150 nm in thickness could be converted into LiF with no indication of MgF$_2$ or Mg impurities in GIXRD, EDX or ToF-ERDA measurements. The lower the reaction temperature, the larger the Lithd dose needed to completely convert the MgF$_2$ film into LiF. Although the resulting LiF films were again highly crystalline as determined with GIXRD, they showed much smaller grain sizes and thus lower roughness than the films deposited with either the two [150] or four [151] step LiF processes (Figure 18). In addition, the adhesion of the films to the silicon substrates was markedly improved. Our experiments demonstrated that the conversion reaction with Lithd is not limited to the surface regions of MgF$_2$ films but can in fact proceed very deeply into the films. The high mobility and reactivity of Li$^+$ seen in these experiments is likely to play a role in many processes used to deposit materials containing lithium, especially in the case of ternaries. For example, Miikkulainen et al. later reported similar results in their conversion experiments to form spinel LiMn$_2$O$_4$ using MnO$_2$ films and Lithd, as was already discussed in a previous section [90]. However, their conversion reaction also led to significant amounts of hydrogen and carbon impurities when no ozone was used after the Lithd pulse, as opposed to our conversion reactions of MgF$_2$ with Lithd which led to purer film products.

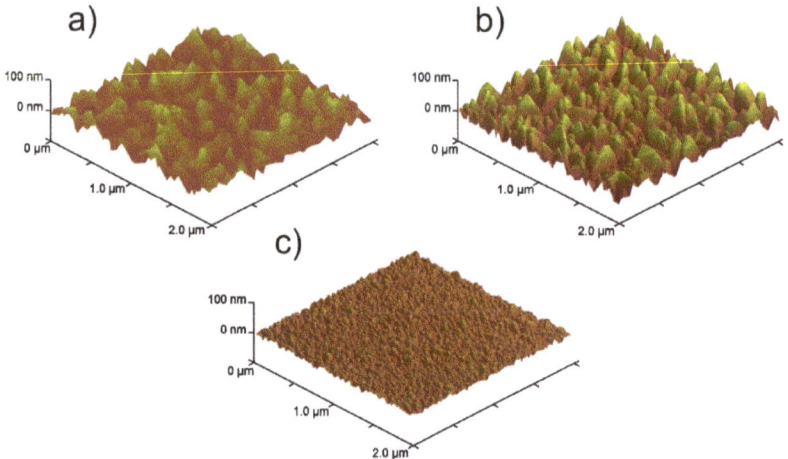

Figure 18. AFM images of LiF films deposited at 325 °C using three different processes: (**a**) Lithd + TiF$_4$, thickness 73 nm, rms roughness 15.9 nm; (**b**) Mg(thd)$_2$ + TiF$_4$ + Lithd + TiF$_4$, thickness 68 nm, rms roughness 20.1 nm; (**c**) conversion from a MgF$_2$ film using Lithd, thickness 94 nm, rms roughness 4.8 nm.

Li$_3$AlF$_6$ was also deposited using a similar conversion reaction as with MgF$_2$ and Lithd (Process 2, Figure 17) [139]. Approximately 50 and 100 nm thin films of AlF$_3$ were exposed to the lithium precursor at different temperatures and for different Lithd pulse numbers to determine whether good quality Li$_3$AlF$_6$ could form from this conversion reaction. GIXRD analyses showed that amorphous AlF$_3$ transformed into monoclinic Li$_3$AlF$_6$ with Lithd exposure. Choosing too large a number of Lithd pulses resulted in crystalline LiF formation. The mechanism of the conversion is most likely similar to the MgF$_2$ conversion, as the oxygen, carbon and hydrogen impurity levels were very low in the films after the conversion, as determined with ToF-ERDA. Despite crystalline Li$_3$AlF$_6$ being visible in the X-ray diffractograms, obtaining the correct Li:Al ratio was challenging. ToF-ERDA revealed that doubling the Lithd exposure from 20 to 40 pulses increased the Li:Al ratio from 0.93:1 to 7.9:1 for approx. 50 nm AlF$_3$ films (Figure 19). With 100 nm AlF$_3$ films exposed to 40 Lithd pulses, the ratio varied between 1.33:1 and 1.49:1. Thus, thinner films were much faster to convert than thicker ones, as was to be expected. The exposure temperature also played a role in the conversion. Just a 25 °C increase from 250 to 275 °C increased the Li:Al ratio for a 40 pulse sample from 1.49:1 to 2:1. However, despite the lithium deficient metal ratio, the 275 °C exposure temperature sample already contained a prominent amount of crystalline LiF based on GIXRD. The converted films showed a porous structure in FESEM (field emission scanning electron microscopy) (Figure 20), preventing ionic conductivity measurements due to top electrodes short circuiting with the bottom electrode.

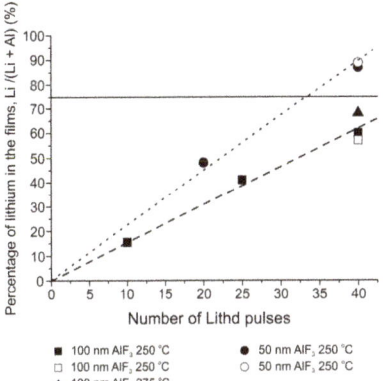

Figure 19. The content of lithium cations in converted AlF$_3$ films as a function of the number of Lithd pulses. Black and white symbols denote samples prepared at different times but with same exposure parameters. Dotted and dashed lines illustrate that the content of lithium increases linearly with the number of Lithd pulses. The solid line illustrates the correct metal stoichiometry of Li$_3$AlF$_6$. Data from [139].

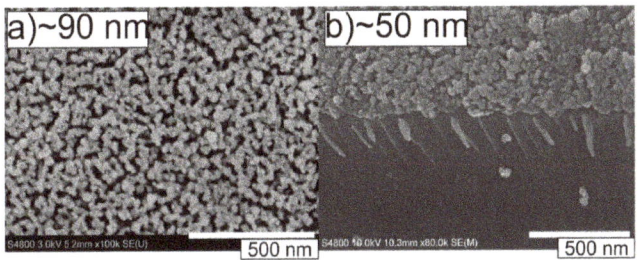

Figure 20. FESEM images of (**a**) ~90 nm crystalline Li$_3$AlF$_6$ film made by conversion from AlF$_3$ and Lithd and (**b**) ~50 nm crystalline Li$_3$AlF$_6$ film made by conversion from AlF$_3$ and Lithd and taken by tilting the sample.

6. Conclusions

Metal fluorides could provide an interesting, high voltage and high capacity alternative to current oxide based lithium-ion battery cathode materials. However, because most metal fluorides act as conversion cathodes, special effort must be made to enhance their cycling ability and to alleviate problems related to electrode pulverization. For some fluorides, encouraging lithium-ion conductivities have been measured, meaning that they could act as highly stable solid electrolytes in all-solid-state Li-ion batteries. In addition, fluoride thin films have been shown to function as artificial solid-electrolyte-interface layers, protecting both cathodes and anodes from metal dissolution and other undesired side reactions during battery cycling. All these results show that there is a lot of untapped potential in this class of battery materials, worthy of more research. For future applications, methods producing high quality, conformal fluoride thin films with precise thickness control are needed, for example for depositing cathodes in intricate 3D structures and for depositing protecting films of 1 nm or less on current electrode materials. Atomic layer deposition can be the method for answering these demands. The lithium-ion exhibits unique ALD chemistries through its high mobility, which needs to be taken into account when designing processes for battery material deposition by ALD.

Many metal fluoride thin films have been deposited using ALD, however the focus of these studies has mainly been on optical applications. Only very recently studies on the deposition of materials such as AlF_3 and L_3AlF_6 have emerged, with the main motivation of utilizing these materials in lithium-ion batteries. In the future, much more research effort should be put into depositing transition metal fluorides and multi-component fluorides for use in lithium-ion batteries. These materials have received next to no interest in the past 25 years of ALD fluoride studies, meaning there is much to be discovered in this area of materials science. These future studies will undoubtedly greatly benefit from recent studies utilizing new fluorine sources such as metal fluorides and SF_6 plasma.

Author Contributions: Writing–Original Draft Preparation, M.M.; Writing–Review & Editing, M.M.; M.R.; M.L.; Visualization, M.M.; Supervision, M.R.; M.L.

Funding: This work was funded by the Finnish Centre of Excellence in Atomic Layer Deposition (284623).

Acknowledgments: Peter J. King is thanked for proof-reading this work.

Conflicts of Interest: The authors declare no conflict of interest.

References

1. Tesla Solar Roof. Available online: https://www.tesla.com/solarroof (accessed on 12 May 2018).
2. Pillot, C. The worldwide battery market 2011–2025. In Proceedings of the Batteries 2012, Nice, France, 24–26 October 2012.
3. Batteries storing power seen as big as rooftop solar in 12 years. Available online: https://www.bloomberg.com/news/articles/2016-06-13/batteries-storing-power-seen-as-big-as-rooftop-solar-in-12-years (accessed on 12 May 2018).
4. Oudenhoven, J.F.M.; Baggetto, L.; Notten, P.H.L. All-solid-state lithium-ion microbatteries: A review of various three-dimensional concepts. *Adv. Energy Mater.* **2011**, *1*, 10–33. [CrossRef]
5. Hayner, C.M.; Zhao, X.; Kung, H.H. Materials for Rechargeable Lithium-Ion Batteries. *Annu. Rev. Chem. Biomol. Eng.* **2012**, *3*, 445–471. [CrossRef] [PubMed]
6. Knoops, H.C.M.; Donders, M.E.; van de Sanden, M.C.M.; Notten, P.H.L.; Kessels, W.M.M. Atomic layer deposition for nanostructured Li-ion batteries. *J. Vac. Sci. Technol. A* **2012**, *30*, 010801. [CrossRef]
7. Marichy, C.; Bechelany, M.; Pinna, N. Atomic Layer Deposition of Nanostructured Materials for Energy and Environmental Applications. *Adv. Mater.* **2012**, *24*, 1017–1032. [CrossRef] [PubMed]
8. Long, J.W.; Dunn, B.; Rolison, D.R.; White, H.S. Three-Dimensional Battery Architectures. *Chem. Rev.* **2004**, *104*, 4463–4492. [CrossRef] [PubMed]
9. Nitta, N.; Yushin, G. High-capacity anode materials for lithium-ion batteries: Choice of elements and structures for active particles. *Part. Part. Syst. Charact.* **2014**, *31*, 317–336. [CrossRef]
10. Zhu, C.; Han, K.; Geng, D.; Ye, H.; Meng, X. Achieving high-performance silicon anodes of lithium-ion batteries via atomic and molecular layer deposited surface coatings: An overview. *Electrochim. Acta* **2017**, *251*, 710–728. [CrossRef]
11. Amatucci, G.G.; Pereira, N. Fluoride based electrode materials for advanced energy storage devices. *J. Fluor. Chem.* **2007**, *128*, 243–262. [CrossRef]
12. Oi, T. Ionic conductivity of amorphous mLiFnMF$_3$ thin films (M = Al, Cr, Sc or Al + Sc). *Mat. Res. Bull.* **1984**, *19*, 1343–1348. [CrossRef]
13. Oi, T. Ionic conductivity of LiF thin films containing Di- or trivalent metal fluorides. *Mat. Res. Bull.* **1984**, *19*, 451–457. [CrossRef]
14. Oi, T. Ionic conductivity of amorphous ternary fluoride thin films of composition $Li_2M^{II}M^{III}F_7$ (M^{II} = Mg, Fe; M^{III} = Al, Sc. *Mat. Res. Bull.* **1984**, *19*, 1077–1082. [CrossRef]
15. Huggins, R.A. *Advanced Batteries: Materials Science Aspects*; Springer: New York, NY, USA, 2009; ISBN 978-0-387-76424-5.
16. Tarascon, J.-M.; Armand, M. Issues and challenges facing rechargeable lithium batteries. *Nature* **2001**, *414*, 359–367. [CrossRef] [PubMed]
17. Baggetto, L.; Niessen, R.A.H.; Roozeboom, F.; Notten, P.H.L. High energy density all-solid-state batteries: A challenging concept towards 3D integration. *Adv. Funct. Mater.* **2008**, *18*, 1057–1066. [CrossRef]

18. Whittingham, M.S. Lithium batteries and cathode materials. *Chem. Rev.* **2004**, *104*, 4271–4302. [CrossRef] [PubMed]
19. Ellis, B.L.; Lee, K.T.; Nazar, L.F. Positive electrode materials for Li-Ion and Li-batteries. *Chem. Mater.* **2010**, *22*, 691–714. [CrossRef]
20. Chen, Z.; Qin, Y.; Amine, K.; Sun, Y.-K. Role of surface coating on cathode materials for lithium-ion batteries. *J. Mater. Chem.* **2010**, *20*, 7606–7612. [CrossRef]
21. Park, J.S.; Meng, X.; Elam, J.W.; Hao, S.; Wolverton, C.; Kim, C.; Cabana, J. Ultrathin lithium-ion conducting coatings for increased interfacial stability in high voltage lithium-ion batteries. *Chem. Mater.* **2014**, *26*, 3128–3134. [CrossRef]
22. Guan, D.; Jeevarajan, J.A.; Wang, Y. Enhanced cycleability of $LiMn_2O_4$ cathodes by atomic layer deposition of nanosized-thin Al_2O_3 coatings. *Nanoscale* **2011**, *3*, 1465–1469. [CrossRef] [PubMed]
23. Jung, Y.S.; Cavanagh, A.S.; Riley, L.A.; Kang, S.-H.; Dillon, A.C.; Groner, M.D.; George, S.M.; Lee, S.-H. Ultrathin direct atomic layer deposition on composite electrodes for highly durable and safe Li-Ion batteries. *Adv. Mater.* **2010**, *22*, 2172–2176. [CrossRef] [PubMed]
24. Jung, Y.S.; Cavanagh, A.S.; Dillon, A.C.; Groner, M.D.; George, S.M.; Lee, S.-H. Enhanced stability of $LiCoO_2$ cathodes in lithium-ion batteries using surface modification by atomic layer deposition. *J. Electrochem. Soc.* **2010**, *157*, A75–A81. [CrossRef]
25. Nitta, N.; Wu, F.; Lee, J.T.; Yushin, G. Li-ion battery materials: Present and future. *Mater. Today* **2015**, *18*, 252–264. [CrossRef]
26. Xu, W.; Wang, J.; Ding, F.; Chen, X.; Nasybulin, E.; Zhang, Y.; Zhang, J.-G. Lithium metal anodes for rechargeable batteries. *Energy Environ. Sci.* **2014**, *7*, 513–537. [CrossRef]
27. Poizot, P.; Laruelle, S.; Grugeon, S.; Dupont, L.; Tarascon, J.-M. Nano-sized transition-metal oxides as negative-electrode materials for lithium-ion batteries. *Nature* **2000**, *407*, 496–499. [CrossRef] [PubMed]
28. Park, C.-M.; Kim, J.-H.; Kim, H.; Sohn, H.-J. Li-alloy based anode materials for Li secondary batteries. *Chem. Soc. Rev.* **2010**, *39*, 3115–3141. [CrossRef] [PubMed]
29. Li, H.; Balaya, P.; Maier, J. Li-Storage via heterogeneous reaction in selected binary metal fluorides and oxides. *J. Electrochem. Soc.* **2004**, *151*, A1878–A1885. [CrossRef]
30. Thangadurai, V.; Weppner, W. Solid state lithium ion conductors: Design considerations by thermodynamic approach. *Ionics* **2002**, *8*, 281–292. [CrossRef]
31. Thangadurai, V.; Weppner, W. Recent progress in solid oxide and lithium ion conducting electrolytes research. *Ionics* **2006**, *12*, 81–92. [CrossRef]
32. Xia, H.; Wang, H.L.; Xiao, W.; Lai, M.O.; Lu, L. Thin film Li electrolytes for all-solid-state micro-batteries. *Int. J. Surf. Sci. Eng.* **2009**, *3*, 23–43. [CrossRef]
33. Knauth, P. Inorganic solid Li ion conductors: An overview. *Solid State Ion.* **2009**, *180*, 911–916. [CrossRef]
34. Fergus, J.W. Ceramic and polymeric solid electrolytes for lithium-ion batteries. *J. Power Sources* **2010**, *195*, 4554–4569. [CrossRef]
35. Hu, M.; Pang, X.; Zhou, Z. Recent progress in high-voltage lithium ion batteries. *J. Power Sources* **2013**, *237*, 229–242. [CrossRef]
36. Cabana, J.; Monconduit, L.; Larcher, D.; Palacín, M.R. Beyond intercalation-based Li-ion batteries: The state of the art and challenges of electrode materials reacting through conversion reactions. *Adv. Mater.* **2010**, *22*, E170–E192. [CrossRef] [PubMed]
37. Li, H.; Richter, G.; Maier, J. Reversible formation and decomposition of LiF clusters using transition metal fluorides as precursors and their application in rechargeable Li batteries. *Adv. Mater.* **2003**, *15*, 736–739. [CrossRef]
38. Badway, F.; Cosandey, F.; Pereira, N.; Amatucci, G.G. Carbon metal fluoride nanocomposites: High-capacity reversible metal fluoride conversion materials as rechargeable positive electrodes for Li batteries. *J. Electrochem. Soc.* **2003**, *150*, A1318–A1327. [CrossRef]
39. Zhang, H.; Zhou, Y.-N.; Sun, Q.; Fu, Z.-W. Nanostructured nickel fluoride thin film as a new Li storage material. *Solid State Sci.* **2008**, *10*, 1166–1172. [CrossRef]
40. Bervas, M.; Badway, F.; Klein, L.C.; Amatucci, G.G. Bismuth fluoride nanocomposite as a positive electrode material for rechargeable lithium batteries. *Electrochem. Solid-State Lett.* **2005**, *8*, A179–A183. [CrossRef]
41. Gonzalo, E.; Kuhn, A.; García-Alvarado, F. On the room temperature synthesis of monoclinic Li_3FeF_6: A new cathode material for rechargeable lithium batteries. *J. Power Sources* **2010**, *195*, 4990–4996. [CrossRef]

42. Basa, A.; Gonzalo, E.; Kuhn, A.; García-Alvarado, F. Reaching the full capacity of the electrode material Li$_3$FeF$_6$ by decreasing the particle size to nanoscale. *J. Power Sources* **2012**, *197*, 260–266. [CrossRef]
43. Basa, A.; Gonzalo, E.; Kuhn, A.; García-Alvarado, F. Facile synthesis of β-Li$_3$VF$_6$: A new electrochemically active lithium insertion material. *J. Power Sources* **2012**, *207*, 160–165. [CrossRef]
44. Koyama, Y.; Tanaka, I.; Adachi, H. New fluoride cathodes for rechargeable lithium batteries. *J. Electrochem. Soc.* **2000**, *147*, 3633–3636. [CrossRef]
45. Amalraj, F.; Talianker, M.; Markovsky, B.; Burlaka, L.; Leifer, N.; Goobes, G.; Erickson, E.M.; Haik, O.; Grinblat, J.; Zinigrad, E.; et al. Studies of Li and Mn-rich Li$_x$[MnNiCo]O$_2$ electrodes: electrochemical performance, structure, and the effect of the aluminum fluoride coating. *J. Electrochem. Soc.* **2013**, *160*, A2220–A2233. [CrossRef]
46. Lee, S.-H.; Yoon, C.S.; Amine, K.; Sun, Y.-K. Improvement of long-term cycling performance of Li[Ni$_{0.8}$Co$_{0.15}$Al$_{0.05}$]O$_2$ by AlF$_3$ coating. *J. Power Sources* **2013**, *234*, 201–207. [CrossRef]
47. Sun, Y.-K.; Lee, M.-J.; Yoon, C.S.; Hassoun, J.; Amine, K.; Scrosati, B. The role of AlF$_3$ coatings in improving electrochemical cycling of Li-enriched nickel-manganese oxide electrodes for Li-ion batteries. *Adv. Mater.* **2012**, *24*, 1192–1196. [CrossRef] [PubMed]
48. Tron, A.; Park, Y.D.; Mun, J. AlF$_3$-coated LiMn$_2$O$_4$ as cathode material for aqueous rechargeable lithium battery with improved cycling stability. *J. Power Sources* **2016**, *325*, 360–364. [CrossRef]
49. Sun, Y.-K.; Cho, S.-W.; Myung, S.-T.; Amine, K.; Prakash, J. Effect of AlF$_3$ coating amount on high voltage cycling performance of LiCoO$_2$. *Electrochim. Acta* **2007**, *53*, 1013–1019. [CrossRef]
50. Ding, F.; Xu, W.; Choi, D.; Wang, W.; Li, X.; Engelhard, M.H.; Chen, X.; Yang, Z.; Zhang, J.-G. Enhanced performance of graphite anode materials by AlF$_3$ coating for lithium-ion batteries. *J. Mater. Chem.* **2012**, *22*, 12745–12751. [CrossRef]
51. Lee, Y.; Piper, D.M.; Cavanagh, A.S.; Young, M.J.; Lee, S.-H.; George, S.M. Atomic layer deposition of lithium ion conducting (AlF$_3$)(LiF)$_x$ alloys using trimethylaluminum, lithium hexamethyldisilazide and hydrogen fluoride-pyridine. In Proceedings of the 14th International Conference on Atomic Layer Deposition, Kyoto, Japan, 15–18 June 2014.
52. Miyazaki, R.; Maekawa, H. Li$^+$-ion conduction of Li$_3$AlF$_6$ mechanically milled with LiCl. *ECS Electrochem. Lett.* **2012**, *1*, A87–A89. [CrossRef]
53. Dance, J.-M.; Oi, T. Ionic conductivity of amorphous lithium fluoride-nickel fluoride (mLiF-nNiF$_2$) thin films. *Thin Solid Films* **1983**, *104*, L71–L73.
54. Kawamoto, Y.; Fujiwara, J.; Ichimura, C. Ionic conduction in xMF·(95 − x)ZrF$_4$·5LaF$_3$ (M: Alkali metals) glasses: I. Lithium ion conduction in xLiF·(95 − x)ZrF$_4$·5LaF$_3$ glasses. *J. Non-Cryst. Solids* **1989**, *111*, 245–251. [CrossRef]
55. Reau, J.M.; Kahnt, H.; Poulain, M. Ionic transport studies in mixed alkali-fluorine conductor glasses of composition ZrF$_4$-BaF$_2$-LaF$_3$-AF (A = Li, Na) and ZrF$_4$-BaF$_2$-ThF$_4$-LiF. *J. Non-Cryst. Solids* **1990**, *119*, 347–350. [CrossRef]
56. Senegas, J.; Reau, J.M.; Aomi, H.; Hagenmuller, P.; Poulain, M. Ionic conductvity and NMR investigation of quaternary glasses in the ZrF$_4$-BaF$_2$-ThF$_4$-LiF system. *J. Non-Cryst. Solids* **1986**, *85*, 315–334. [CrossRef]
57. Trnovcová, V.; Fedorov, P.P.; Bárta, Č.; Labaš, V.; Meleshina, V.A.; Sobolev, B.P. Microstructure and physical properties of superionic eutectic composites of the LiF-RF$_3$ (R = rare earth element) system. *Solid State Ion.* **1999**, *119*, 173–180. [CrossRef]
58. Trnovcová, V.; Fedorov, P.P.; Furár, I. Fluoride solid electrolytes containing rare earth elements. *J. Rare Earths* **2008**, *26*, 225–232. [CrossRef]
59. Dieudonné, B.; Chable, J.; Mauvy, F.; Fourcade, S.; Durand, E.; Lebraud, E.; Leblanc, M.; Legein, C.; Body, M.; Maisonneuve, V.; et al. Exploring the Sm$_{1−x}$Ca$_x$F$_{3−x}$ tysonite solid solution as a solid-state electrolyte: Relationships between structural features and F$^−$ ionic conductivity. *J. Phys. Chem. C* **2015**, *119*, 25170–25179. [CrossRef]
60. Sorokin, N.I.; Sobolev, B.P. Nonstoichiometric fluorides-solid electrolytes for electrochemical devices: A review. *Crystallogr. Rep.* **2007**, *52*, 842–863. [CrossRef]
61. Reddy, M.A.; Fichtner, M. Batteries based on fluoride shuttle. *J. Mater. Chem.* **2011**, *21*, 17059–17062. [CrossRef]
62. Leskelä, M.; Ritala, M. Atomic layer deposition chemistry: Recent developments and future challenge. *Angew. Chem. Int. Ed.* **2003**, *42*, 5548–5554. [CrossRef] [PubMed]

63. George, S.M. Atomic layer deposition: An overview. *Chem. Rev.* **2010**, *110*, 111–131. [CrossRef] [PubMed]
64. Ritala, M.; Leskelä, M. Atomic layer deposition. In *Handbook of Thin Film Materials*; Nalwa, H.S., Ed.; Academic Press: San Diego, CA, USA, 2002; Volume 1, pp. 103–159. ISBN 9780125129084.
65. Profijt, H.B.; Potts, S.E.; van de Sanden, M.C.M.; Kessels, W.M.M. Plasma-assisted atomic layer deposition: Basics, opportunities, and challenges. *J. Vac. Sci. Technol. A* **2011**, *29*, 050801. [CrossRef]
66. Chalker, P.R. Photochemical atomic layer deposition and etching. *Surf. Coat. Technol.* **2016**, *291*, 258–263. [CrossRef]
67. Miikkulainen, V.; Väyrynen, K.; Kilpi, V.; Han, Z.; Vehkamäki, M.; Mizohata, K.; Räisänen, J.; Ritala, M. Photo-assisted ALD: Process development and application perspectives. *ECS Trans.* **2017**, *80*, 49–60. [CrossRef]
68. Puurunen, R.L. Growth per cycle in atomic layer deposition: A theoretical model. *Chem. Vap. Depos.* **2003**, *9*, 249–257. [CrossRef]
69. Dunn, B.; Long, J.W.; Rolison, D.R. Rethinking multifunction in three dimensions for miniaturizing electrical energy storage. *Electrochem. Soc. Interface* **2008**, *17*, 49–53.
70. Putkonen, M.; Aaltonen, T.; Alnes, M.; Sajavaara, T.; Nilsen, O.; Fjellvåg, H. Atomic layer deposition of lithium containing thin films. *J. Mater. Chem.* **2009**, *19*, 8767–8771. [CrossRef]
71. Nilsen, O.; Miikkulainen, V.; Gandrud, K.B.; Østreng, E.; Ruud, A.; Fjellvåg, H. Atomic layer deposition of functional films for Li-ion microbatteries. *Phys. Status Solidi A* **2014**, *211*, 357–367. [CrossRef]
72. Aaltonen, T.; Miikkulainen, V.; Gandrud, K.B.; Pettersen, A.; Nilsen, O.; Fjellvåg, H. ALD of Thin Films for Lithium-Ion Batteries. *ECS Trans.* **2011**, *41*, 331–339. [CrossRef]
73. Meng, X.; Wang, X.; Geng, D.; Ozgit-Akgun, C.; Schneider, N.; Elam, J.W. Atomic layer deposition for nanomaterial synthesis and functionalization in energy technology. *Mater. Horiz.* **2017**, *4*, 133–154. [CrossRef]
74. Meng, X.; Elam, J.W. Atomic layer deposition of nanophase materials for electrical energy storage. *ECS Trans.* **2015**, *69*, 39–57. [CrossRef]
75. Meng, X.; Yang, X.-Q.; Sun, X. Emerging applications of atomic layer deposition for lithium-ion battery studies. *Adv. Mater.* **2012**, *24*, 3589–3615. [CrossRef] [PubMed]
76. Guan, C.; Wang, J. Recent development of advanced electrode materials by atomic layer deposition for electrochemical energy storage. *Adv. Sci.* **2016**, *3*, 1500405. [CrossRef] [PubMed]
77. Ma, L.; Nuwayhid, R.B.; Wu, T.; Lei, Y.; Amine, K.; Lu, J. Atomic layer deposition for lithium-based batteries. *Adv. Mater. Interfaces* **2016**, *3*, 1600564. [CrossRef]
78. Nilsen, O.; Gandrud, K.B.; Ruud, A.; Fjellvåg, H. Atomic layer deposition for thin-film lithium-ion batteries. In *Atomic Layer Deposition in Energy Conversion Applications*; Bachmann, J., Ed.; Wiley-VCH Verlag GmbH & Co. KGaA: Weinheim, Germany, 2017; pp. 183–207. ISBN 978-3-527-33912-9.
79. Mäntymäki, M. Atomic Layer Deposition and Lithium-ion Batteries: Studies on New Materials and Reactions for Battery Development. Ph.D. Thesis, University of Helsinki, Helsinki, Finland, 9 June 2017.
80. Liu, J.; Sun, X. Elegant design of electrode and electrode/electrolyte interface in lithium-ion batteries by atomic layer deposition. *Nanotechnology* **2015**, *26*, 024001. [CrossRef] [PubMed]
81. Wang, X.; Yushin, G. Chemical vapor deposition and atomic layer deposition for advanced lithium ion batteries and supercapacitors. *Energy Environ. Sci.* **2015**, *8*, 1889–1904. [CrossRef]
82. Noked, M.; Liu, C.; Hu, J.; Gregorczyk, K.; Rubloff, G.W.; Lee, S.B. Electrochemical thin layers in nanostructures for energy storage. *Acc. Chem. Res.* **2016**, *49*, 2336–2346. [CrossRef] [PubMed]
83. Share, K.; Westover, A.; Li, M.; Pint, C.L. Surface engineering of nanomaterials for improved energy storage —A review. *Chem. Eng. Sci.* **2016**, *154*, 3–19. [CrossRef]
84. Sun, Q.; Lau, K.C.; Geng, D.; Meng, X. Atomic and molecular layer deposition for superior lithium-sulfur batteries: Strategies, performance, and mechanisms. *Batter. Supercaps* **2018**. [CrossRef]
85. Le Van, K.; Groult, H.; Mantoux, A.; Perrigaud, L.; Lantelme, F.; Lindström, R.; Badour-Hadjean, R.; Zanna, S.; Lincot, D. Amorphous vanadium oxide films synthesised by ALCVD for lithium rechargeable batteries. *J. Power Sources* **2006**, *160*, 592–601. [CrossRef]
86. Chen, X.; Pomerantseva, E.; Gregorczyk, K.; Ghodssi, R.; Rubloff, G. Cathodic ALD V_2O_5 thin films for high-rate electrochemical energy storage. *RSC Adv.* **2013**, *3*, 4294–4302. [CrossRef]
87. Donders, M.E.; Knoops, H.C.M.; Kessels, W.M.M.; Notten, P.H.L. Remote plasma atomic layer deposition of thin films of electrochemically active $LiCoO_2$. *ECS Trans.* **2011**, *41*, 321–330. [CrossRef]

88. Donders, M.E.; Arnoldbik, W.M.; Knoops, H.C.M.; Kessels, W.M.M.; Notten, P.H.L. Atomic layer deposition of LiCoO$_2$ thin-film electrodes for all-solid-state Li-ion micro-batteries. *J. Electrochem. Soc.* **2013**, *160*, A3066–A3071. [CrossRef]
89. Liu, J.; Banis, M.N.; Sun, Q.; Lushington, A.; Li, R.; Sham, T.-K.; Sun, X. Rational design of atomic-layer-deposited LiFePO$_4$ as a high-performance cathode for lithium-ion batteries. *Adv. Mater.* **2014**, *26*, 6472–6477. [CrossRef] [PubMed]
90. Miikkulainen, V.; Ruud, A.; Østreng, E.; Nilsen, O.; Laitinen, M.; Sajavaara, T.; Fjellvåg, H. Atomic layer deposition of spinel lithium manganese oxide by film-body-controlled lithium incorporation for thin-film lithium-ion batteries. *J. Phys. Chem. C* **2014**, *118*, 1258–1268. [CrossRef]
91. Meng, X.; Comstock, D.J.; Fisher, T.T.; Elam, J.W. Vapor-phase atomic-controllable growth of amorphous Li$_2$S for high-performance lithium–sulfur batteries. *ACS Nano* **2014**, *8*, 10963–10972. [CrossRef] [PubMed]
92. Lantelme, F.; Mantoux, A.; Groult, H.; Lincot, D. Electrochemical study of phase transition processes in lithium insertion in V$_2$O$_5$ electrodes. *J. Electrochem. Soc.* **2003**, *150*, A1202–A1208. [CrossRef]
93. Gandrud, K.B.; Pettersen, A.; Nilsen, O.; Fjellvåg, H. Growth of LiFePO$_4$ cathode material by ALD. In Proceedings of the 10th International Conference on Atomic Layer Deposition, Seoul, Korea, 20–23 June 2010.
94. Gandrud, K.B.; Pettersen, A.; Nilsen, O.; Fjellvåg, H. High-performing iron phosphate for enhanced lithium ion solid state batteries as grown by atomic layer deposition. *J. Mater. Chem. A* **2013**, *1*, 9054–9059. [CrossRef]
95. Kozen, A.C.; Pearse, A.J.; Lin, C.-F.; Schoeder, M.A.; Noked, M.; Lee, S.B.; Rubloff, G.W. Atomic layer deposition and in situ characterization of ultraclean lithium oxide and lithium hydroxide. *J. Phys. Chem. C* **2014**, *118*, 27749–27753. [CrossRef]
96. Wang, W.; Tian, M.; Abdulagatov, A.; George, S.M.; Lee, Y.-C.; Yang, R. Three-dimensional Ni/TiO$_2$ nanowire network for high areal capacity lithium ion microbattery applications. *Nano Lett.* **2012**, *12*, 655–660. [CrossRef] [PubMed]
97. Cheah, S.K.; Perre, E.; Rooth, M.; Fondell, M.; Hårsta, A.; Nyholm, L.; Boman, M.; Gustafsson, T.; Lu, J.; Simon, P.; et al. Self-supported three-dimensional nanoelectrodes for microbattery applications. *Nano Lett.* **2009**, *9*, 3230–3233. [CrossRef] [PubMed]
98. Panda, S.K.; Yoon, Y.; Jung, H.S.; Yoon, W.-S.; Shin, H. Nanoscale size effect of titania (anatase) nanotubes with uniform wall thickness as high performance anode for lithium-ion secondary battery. *J. Power Sources* **2012**, *204*, 162–167. [CrossRef]
99. Miikkulainen, V.; Nilsen, O.; Laitinen, M.; Sajavaara, T.; Fjellvåg, H. Atomic layer deposition of Li$_x$Ti$_y$O$_z$ thin films. *RSC Adv.* **2013**, *3*, 7537–7542. [CrossRef]
100. Meng, X.; Liu, J.; Li, X.; Banis, M.N.; Yang, J.; Li, R.; Sun, X. Atomic layer deposited Li$_4$Ti$_5$O$_{12}$ on nitrogen-doped carbon nanotubes. *RSC Adv.* **2013**, *3*, 7285–7288. [CrossRef]
101. Miikkulainen, V.; Nilsen, O.; Laitinen, M.; Sajavaara, T.; Fjellvåg, H. Atomic layer deposition of Li$_x$Ti$_y$O$_z$ films. In Proceedings of the 12th International Conference on Atomic Layer Deposition, Dresden, Germany, 17–20 June 2012.
102. Nisula, M.; Karppinen, M. Atomic/molecular layer deposition of lithium terephthalate thin films as high rate capability Li-Ion battery anodes. *Nano Lett.* **2016**, *16*, 1276–1281. [CrossRef] [PubMed]
103. Shen, Y.; Søndergaard, M.; Christensen, M.; Birgisson, S.; Iversen, B.B. Solid state formation mechanism of Li$_4$Ti$_5$O$_{12}$ from an anatase TiO$_2$ source. *Chem. Mater.* **2014**, *26*, 3679–3686. [CrossRef]
104. Atosuo, E.; Mäntymäki, M.; Mizohata, K.; Heikkilä, M.J.; Räisänen, J.; Ritala, M.; Leskelä, M. Preparation of lithium containing oxides by the solid state reaction of atomic layer deposited thin films. *Chem. Mater.* **2017**, *29*, 998–1005. [CrossRef]
105. Sundberg, P.; Karppinen, M. Organic and inorganic–organic thin film structures by molecular layer deposition: A review. *Beilstein J. Nanotechnol.* **2014**, *5*, 1104–1136. [CrossRef] [PubMed]
106. Armand, M.; Grugeon, S.; Vezin, H.; Laruelle, S.; Ribière, P.; Poizot, P.; Tarascon, J.-M. Conjugated dicarboxylate anodes for Li-ion batteries. *Nat. Mater.* **2009**, *8*, 120–125. [CrossRef] [PubMed]
107. Nisula, M.; Karppinen, M. In situ lithiated quinone cathode for ALD/MLD-fabricated high-power thin-film battery. *J. Mater. Chem. A* **2018**, *6*, 7027–7033. [CrossRef]
108. Nisula, M.; Shindo, Y.; Koga, H.; Karppinen, M. Atomic layer deposition of lithium phosphorus oxynitride. *Chem. Mater.* **2015**, *27*, 6987–6993. [CrossRef]

109. Kozen, A.C.; Pearse, A.J.; Lin, C.-F.; Noked, M.; Rubloff, G.W. Atomic layer deposition of the solid electrolyte LiPON. *Chem. Mater.* **2015**, *27*, 5324–5331. [CrossRef]
110. Hämäläinen, J.; Munnik, F.; Hatanpää, T.; Holopainen, J.; Ritala, M.; Leskelä, M. Study of amorphous lithium silicate thin films grown by atomic layer deposition. *J. Vac. Sci. Technol. A* **2012**, *30*, 01A106. [CrossRef]
111. Tomczak, Y.; Knapas, K.; Sundberg, M.; Leskelä, M.; Ritala, M. In situ reaction mechanism studies on lithium hexadimethyldisilazide and ozone atomic layer deposition process for lithium silicate. *J. Phys. Chem. C* **2013**, *117*, 14241–14246. [CrossRef]
112. Ruud, A.; Miikkulainen, V.; Mizohata, K.; Fjellvåg, H.; Nilsen, O. Enhanced process and composition control for atomic layer deposition with lithium trimethylsilanolate. *J. Vac. Sci. Technol. A* **2017**, *35*, 01B133-1–01B133-8. [CrossRef]
113. Wang, B.; Liu, J.; Banis, M.N.; Sun, Q.; Zhao, Y.; Li, R.; Sham, T.-K.; Sun, X. Atomic layer deposited lithium silicates as solid-state electrolytes for all-solid-state batteries. *ACS Appl. Mater. Interfaces* **2017**, *9*, 31786–31793. [CrossRef] [PubMed]
114. Hämäläinen, J.; Holopainen, J.; Munnik, F.; Hatanpää, T.; Heikkilä, M.; Ritala, M.; Leskelä, M. Lithium phosphate thin films grown by atomic layer deposition. *J. Electrochem. Soc.* **2012**, *159*, A259–A263. [CrossRef]
115. Wang, B.; Liu, J.; Sun, Q.; Li, R.; Sham, T.-K.; Sun, X. Atomic layer deposition of lithium phosphates as solid-state electrolytes for all-solid-state microbatteries. *Nanotechnology* **2014**, *25*, 504007. [CrossRef] [PubMed]
116. Létiche, M.; Eustache, E.; Freixas, J.; Demortière, A.; De Andrade, V.; Morgenroth, L.; Tilmant, P.; Vaurette, F.; Troadec, D.; Roussel, P.; et al. Atomic layer deposition of functional layers for on chip 3D Li-ion all solid state microbattery. *Adv. Energy Mater.* **2017**, *7*, 1601402. [CrossRef]
117. Liu, J.; Banis, M.N.; Li, X.; Lushington, A.; Cai, M.; Li, R.; Sham, T.-K.; Sun, X. Atomic layer deposition of lithium tantalate solid-state electrolytes. *J. Phys. Chem. C* **2013**, *117*, 20260–20267. [CrossRef]
118. Østreng, E.; Sønsteby, H.H.; Sajavaara, T.; Nilsen, O.; Fjellvåg, H. Atomic layer deposition of ferroelectric LiNbO$_3$. *J. Mater. Chem. C* **2013**, *1*, 4283–4290. [CrossRef]
119. Wang, B.; Zhao, Y.; Banis, M.N.; Sun, Q.; Adair, K.R.; Li, R.; Sham, T.-K.; Sun, Q. Atomic layer deposition of lithium niobium oxides as potential solid-state electrolytes for lithium-ion batteries. *ACS Appl. Mater. Interfaces* **2018**, *10*, 1654–1661. [CrossRef] [PubMed]
120. Aaltonen, T.; Alnes, M.; Nilsen, O.; Costelle, L.; Fjellvåg, H. Lanthanum titanate and lithium lanthanum titanate thin films grown by atomic layer deposition. *J. Mater. Chem.* **2010**, *20*, 2877–2881. [CrossRef]
121. Perng, Y.-C.; Cho, J.; Sun, S.Y.; Membreno, D.; Cirigliano, N.; Dunn, B.; Chang, J.P. Synthesis of ion conducting Li$_x$Al$_y$Si$_z$O thin films by atomic layer deposition. *J. Mater. Chem. A* **2014**, *2*, 9566–9573. [CrossRef]
122. Kazyak, E.; Chen, K.-H.; Wood, K.N.; Davis, A.L.; Thompson, T.; Bielinski, A.R.; Sanchez, A.J.; Wang, X.; Wang, C.; Sakamoto, J.; et al. Atomic layer deposition of the solid electrolyte garnet Li$_7$La$_3$Zr$_2$O$_{12}$. *Chem. Mater.* **2017**, *29*, 3785–3792. [CrossRef]
123. Cao, Y.; Meng, X.; Elam, J.W. Atomic Layer Deposition of Li$_x$Al$_y$S Solid-state electrolytes for stabilizing lithium-metal anodes. *ChemElectroChem* **2016**, *3*, 858–863. [CrossRef]
124. Xie, J.; Sendek, A.D.; Cubuk, E.D.; Zhang, X.; Lu, Z.; Gong, Y.; Wu, T.; Shi, F.; Liu, W.; Reed, E.J.; et al. Atomic layer deposition of stable LiAlF$_4$ lithium ion conductive interfacial layer for stable cathode cycling. *ACS Nano* **2017**, *11*, 7019–7027. [CrossRef] [PubMed]
125. Nakagawa, A.; Kuwata, N.; Matsuda, Y.; Kawamura, J. Characterization of stable solid electrolyte lithium silicate for thin film lithium battery. *J. Phys. Soc. Jpn. Suppl. A* **2010**, *79*, 98–101. [CrossRef]
126. Furusawa, S.; Kamiyama, A.; Tsurui, T. Fabrication and ionic conductivity of amorphous lithium meta-silicate thin film. *Solid State Ion.* **2008**, *179*, 536–542. [CrossRef]
127. Furusawa, S.; Kasahara, T.; Kamiyama, A. Fabrication and ionic conductivity of Li$_2$SiO$_3$ thin film. *Solid State Ion.* **2009**, *180*, 649–653. [CrossRef]
128. Liu, J.; Wang, B.; Sun, Q.; Li, R.; Sham, T.-K.; Sun, X. Atomic layer deposition of hierarchical CNTs@FePO$_4$ architecture as a 3D electrode for lithium-ion and sodium-ion batteries. *Adv. Mater. Interfaces* **2016**, *3*, 1600468. [CrossRef]
129. Wang, B.; Liu, J.; Sun, Q.; Xiao, B.; Li, R.; Sham, T.-K.; Sun, X. Titanium dioxide/lithium phosphate nanocomposite derived from atomic layer deposition as a high-performance anode for lithium ion batteries. *Adv. Mater. Interfaces* **2016**, *3*, 1600369. [CrossRef]

130. Shibata, S. Thermal atomic layer deposition of lithium phosphorus oxynitride as a thin-film solid electrolyte. *J. Electrochem. Soc.* **2016**, *163*, A2555–A2562. [CrossRef]
131. Lin, C.-F.; Noked, M.; Kozen, A.C.; Liu, C.; Zhao, O.; Gregorczyk, K.; Hu, L.; Lee, S.B.; Rubloff, G.W. Solid Electrolyte Lithium Phosphous Oxynitride as a Protective Nanocladding Layer for 3D high-capacity conversion electrodes. *ACS Nano* **2016**, *10*, 2693–2701. [CrossRef] [PubMed]
132. Smith, R.T. Elastic, piezoelectric, and dielectric properties of lithium tantalate. *Appl. Phys. Lett.* **1967**, *11*, 146–148. [CrossRef]
133. Tomeno, I.; Matsumura, S. Dielectric properties of $LiTaO_3$. *Phys. Rev. B* **1988**, *38*, 606–614. [CrossRef]
134. Glass, A.M.; Nassau, K.; Negran, T.J. Ionic conductivity of quenched alkali niobate and tantalate glasses. *J. Appl. Phys.* **1978**, *49*, 4808–4811. [CrossRef]
135. Li, X.; Liu, J.; Banis, M.N.; Lushington, A.; Li, R.; Cai, M.; Sun, X. Atomic layer deposition of solid-state electrolyte coated cathode materials with superior high-voltage cycling behavior for lithium ion battery application. *Energy Environ. Sci.* **2014**, *7*, 768–778. [CrossRef]
136. Comstock, D.; Elam, J.W. Mechanistic study of lithium aluminum oxide atomic layer deposition. *J. Phys. Chem. C* **2013**, *117*, 1677–1683. [CrossRef]
137. Murugan, R.; Thangadurai, V.; Weppner, W. Fast lithium ion conduction in garnet-type $Li_7La_3Zr_2O_{12}$. *Angew. Chem. Int. Ed.* **2007**, *46*, 7778–7781. [CrossRef] [PubMed]
138. Meng, X.; Cao, Y.; Libera, J.A.; Elam, J.W. Atomic layer deposition of aluminum sulfide: growth mechanism and electrochemical evaluation in lithium-ion batteries. *Chem. Mater.* **2017**, *29*, 9043–9052. [CrossRef]
139. Mäntymäki, M.; Mizohata, K.; Heikkilä, M.J.; Räisänen, J.; Ritala, M.; Leskelä, M. Studies on Li_3AlF_6 thin film deposition utilizing conversion reactions of thin films. *Thin Solid Films* **2017**, *636*, 26–33. [CrossRef]
140. Nykänen, E.; Soininen, P.; Niinistö, L.; Leskelä, M.; Rauhala, E. Electroluminescent SrS:Ce,F thin films deposited by the atomic layer epitaxy process. In Proceedings of the 1994 International Workshop on Electroluminescence, Beijing, China, 10–12 October 1994; pp. 437–444.
141. Ylilammi, M.; Ranta-aho, T. Metal fluoride thin films prepared by atomic layer deposition. *J. Electrochem. Soc.* **1994**, *141*, 1278–1284. [CrossRef]
142. Pilvi, T. Atomic Layer Deposition for Optical Applications: Metal fluoride thin films and novel devices. Ph.D. Thesis, University of Helsinki, Helsinki, Finland, 5 December 2008.
143. Lee, Y.; DuMont, J.W.; Cavanagh, A.S.; George, S.M. Atomic Layer Deposition of AlF_3 using trimethylaluminum and hydrogen fluoride. *J. Phys. Chem. C* **2015**, *119*, 14185–14194. [CrossRef]
144. Park, J.S.; Mane, A.U.; Elam, J.W.; Croy, J.R. Amorphous metal fluoride passivation coatings prepared by atomic layer deposition on $LiCoO_2$ for Li-ion batteries. *Chem. Mater.* **2015**, *27*, 1917–1920. [CrossRef]
145. Jackson, D.H.K.; Laskar, M.R.; Fang, S.; Xu, S.; Ellis, R.G.; Li, X.; Dreibelbis, M.; Babcock, S.E.; Mahanthappa, M.K.; Morgan, D.; et al. Optimizing AlF_3 atomic layer deposition using trimethylaluminum and TaF_5: Application to high voltage Li-ion battery cathodes. *J. Vac. Sci. Technol. A* **2016**, *34*, 031503-1–031503-8. [CrossRef]
146. Lee, Y. Atomic layer etching of metal oxides and atomic layer deposition of metal fluorides. Ph.D. Thesis, University of Colorado, Boulder, CO, USA, January 2015.
147. Lee, Y.; Sun, H.; Young, M.J.; George, S.M. Atomic layer deposition of metal fluorides using HF–pyridine as the fluorine precursor. *Chem. Mater.* **2016**, *28*, 2022–2032. [CrossRef]
148. Hennessy, J.; Jewell, A.P.; Greer, F.; Lee, M.C.; Nikzad, S. Atomic layer deposition of magnesium fluoride via bis(ethylcyclopentadienyl)magnesium and anhydrous hydrogen fluoride. *J. Vac. Sci. Technol. A* **2015**, *33*, 01A125. [CrossRef]
149. Hennessy, J.; Jewell, A.D.; Balasubramanian, K.; Nikzad, S. Ultraviolet optical properties of aluminum fluoride thin films deposited by atomic layer deposition. *J. Vac. Sci. Technol. A* **2016**, *34*, 01A120. [CrossRef]
150. Mäntymäki, M.; Hämäläinen, J.; Puukilainen, E.; Munnik, F.; Ritala, M.; Leskelä, M. Atomic layer deposition of LiF thin films from lithd and TiF_4 precursors. *Chem. Vap. Depos.* **2013**, *19*, 111–116. [CrossRef]
151. Mäntymäki, M.; Hämäläinen, J.; Puukilainen, E.; Sajavaara, T.; Ritala, M.; Leskelä, M. Atomic layer deposition of LiF thin films from Lithd, $Mg(thd)_2$, and TiF_4 precursors. *Chem. Mater.* **2013**, *25*, 1656–1663. [CrossRef]
152. Mane, A.; Libera, J.; Elam, J. Atomic layer deposition of LiF thin films using lithium tert-butoxide and metal fluoride precursors. In Proceedings of the 16th International Conference on Atomic Layer Deposition, Dublin, Ireland, 24–27 July 2016.

153. Pilvi, T.; Hatanpää, T.; Puukilainen, E.; Arstila, K.; Bischoff, M.; Kaiser, U.; Kaiser, N.; Leskelä, M.; Ritala, M. Study of a novel ALD process for depositing MgF_2 thin films. *J. Mater. Chem.* **2007**, *17*, 5077–5083. [CrossRef]
154. Pilvi, T.; Puukilainen, E.; Kreissig, U.; Leskelä, M.; Ritala, M. Atomic layer deposition of MgF_2 thin films using TaF_5 as a novel fluorine source. *Chem. Mater.* **2008**, *20*, 5023–5028. [CrossRef]
155. Pilvi, T.; Arstila, K.; Leskelä, M.; Ritala, M. Novel ALD process for depositing CaF_2 thin films. *Chem. Mater.* **2007**, *19*, 3387–3392. [CrossRef]
156. Mäntymäki, M.; Heikkilä, M.J.; Puukilainen, E.; Mizohata, K.; Marchand, B.; Räisänen, J.; Ritala, M.; Leskelä, M. Atomic layer deposition of AlF_3 thin films using halide precursors. *Chem. Mater.* **2015**, *27*, 604–611. [CrossRef]
157. Pilvi, T.; Puukilainen, E.; Munnik, F.; Leskelä, M.; Ritala, M. ALD of YF_3 thin films from TiF_4 and $Y(thd)_3$ precursors. *Chem. Vap. Depos.* **2009**, *15*, 27–32. [CrossRef]
158. Pilvi, T.; Puukilainen, E.; Arstila, K.; Leskelä, M.; Ritala, M. Atomic layer deposition of LaF_3 thin films using $La(thd)_3$ and TiF_4 as precursors. *Chem. Vap. Depos.* **2008**, *14*, 85–91. [CrossRef]
159. Park, J.S.; Mane, A.U.; Elam, J.W.; Croy, J.R. Atomic layer deposition of Al–W–Fluoride on $LiCoO_2$ cathodes: Comparison of particle- and electrode-level coatings. *ACS Omega* **2017**, *2*, 3724–3729. [CrossRef]
160. Putkonen, M.; Szeghalmi, A.; Pippel, E.; Knez, M. Atomic layer deposition of metal fluorides through oxide chemistry. *J. Mater. Chem.* **2011**, *21*, 14461–14465. [CrossRef]
161. Vos, M.F.J.; Knoops, H.C.M.; Synowicki, R.A.; Kessels, W.M.M.; Mackus, A.J.M. Atomic layer deposition of aluminum fluoride using $Al(CH_3)_3$ and SF_6 plasma. *Appl. Phys. Lett.* **2017**, *111*, 113105-1–113105-5. [CrossRef]
162. Bridou, F.; Cuniot-Ponsard, M.; Desvignes, J.-M.; Richter, M.; Kroth, U.; Gottwald, A. Experimental determination of optical constants of MgF_2 and AlF_3 thin films in the vacuum ultra-violet wavelength region (60–124 nm), and its application to optical designs. *Opt. Commun.* **2010**, *283*, 1351–1358. [CrossRef]
163. Sun, J.; Li, X.; Zhang, W.; Yi, K.; Shao, J. Effects of substrate temperatures and deposition rates on properties of aluminum fluoride thin films in deep-ultraviolet region. *Appl. Opt.* **2012**, *51*, 8481–8489. [CrossRef] [PubMed]
164. König, D.; Scholz, R.; Zahn, D.R.T.; Ebest, G. Band diagram of the $AlF_3/SiO_2/Si$ system. *J. Appl. Phys.* **2005**, *97*, 093707-1–093707-9. [CrossRef]
165. Song, G.-M.; Wu, Y.; Liu, G.; Xu, Q. Influence of AlF_3 coating on the electrochemical properties of $LiFePO_4$/graphite Li-ion batteries. *J. Alloy. Compd.* **2009**, *487*, 214–217. [CrossRef]
166. Lee, D.-J.; Lee, K.-S.; Myung, S.-T.; Yashiro, H.; Sun, Y.-K. Improvement of electrochemical properties of $Li_{1.1}Al_{0.05}Mn_{1.85}O_4$ achieved by an AlF_3 coating. *J. Power Sources* **2011**, *196*, 1353–1357. [CrossRef]
167. Lee, H.J.; Kim, S.B.; Park, Y.J. Enhanced electrochemical properties of fluoride-coated $LiCoO_2$ thin films. *Nanoscale Res. Lett.* **2012**, *7*, 16. [CrossRef] [PubMed]
168. Lapiano-Smith, D.A.; Eklund, E.A.; Himpsel, F.J.; Terminello, L.J. Epitaxy of LiF on Ge(100). *Appl. Phys. Lett.* **1991**, *59*, 2174–2176. [CrossRef]
169. Li, H.H. Refractive index of alkali halides and its wavelength and temperature derivatives. *J. Phys. Chem. Ref. Data* **1976**, *5*, 329–528. [CrossRef]
170. Lee, Y.; Huffman, C.; George, S.M. Selectivity in thermal atomic layer etching using sequential, self-limiting fluorination and ligand-exchange reactions. *Chem. Mater.* **2016**, *28*, 7657–7665. [CrossRef]
171. Zywotko, D.R.; George, S.M. Thermal atomic layer etching of ZnO by a "Conversion-Etch" mechanism using sequential exposures of hydrogen fluoride and trimethylaluminum. *Chem. Mater.* **2017**, *29*, 1183–1191. [CrossRef]
172. DuMont, J.W.; George, S.M. Competition between Al_2O_3 atomic layer etching and AlF_3 atomic layer deposition using sequential exposures of trimethylaluminum and hydrogen fluoride. *J. Chem. Phys.* **2017**, *146*, 052819. [CrossRef] [PubMed]
173. Zhou, Y.; Lee, Y.; Sun, H.; Wallas, J.M.; George, S.M.; Xie, M. Coating solution for high-voltage cathode: AlF_3 atomic layer deposition for freestanding $LiCoO_2$ electrodes with high energy density and excellent flexibility. *ACS Appl. Mater. Interfaces* **2017**, *9*, 9614–9619. [CrossRef] [PubMed]
174. Rousseau, F.; Jain, A.; Kodas, T.T.; Hampden-Smith, M.; Farr, J.D.; Muenchausen, R. Low-temperature dry etching of metal oxides and ZnS via formation of volatile metal β-diketonate complexes. *J. Mater. Chem.* **1992**, *2*, 893–894. [CrossRef]

175. Ritala, M.; Leskelä, M.; Nykänen, E.; Soininen, P.; Niinistö, L. Growth of titanium dioxide thin films by atomic layer epitaxy. *Thin Solid Films* **1993**, *225*, 288–295. [CrossRef]
176. Miikkulainen, V.; Leskelä, M.; Ritala, M.; Puurunen, R.L. Crystallinity of inorganic films grown by atomic layer deposition: Overview and general trends. *J. Appl. Phys.* **2013**, *113*, 021301-1–021301-101. [CrossRef]
177. Krause, R.F., Jr.; Douglas, T.B. Heats of formation of $AlClF_2$ and $AlCl_2F$ from subliming aluminum fluoride in the presence of aluminum chloride vapor. *J. Phys. Chem.* **1968**, *72*, 3444–3451. [CrossRef]
178. Haukka, S. ALD technology—Present and future challenges. *ECS Trans.* **2007**, *3*, 15–26. [CrossRef]
179. Klug, J.A.; Proslier, T.; Elam, J.W.; Cook, R.E.; Hiller, J.M.; Claus, H.; Becker, N.G.; Pellin, M.J. Atomic layer deposition of amorphous niobium carbide-based thin film superconductors. *J. Phys. Chem. C* **2011**, *115*, 25063–25071. [CrossRef]
180. Kaipio, M.; Kemell, M.; Vehkamäki, M.; Mattinen, M.; Mizohata, K.; Ritala, M.; Leskelä, M. Atomic layer deposition of metal carbides—The $TiCl_4$/TMA process as an example. In Proceedings of the 14th Baltic Conference on Atomic Layer Deposition, St. Petersburg, Russia, 2–4 October 2016.
181. Chen, Y.; Ould-Chikh, S.; Abou-Hamad, E.; Callens, E.; Mohandas, J.C.; Khalid, S.; Basset, J.-M. Facile and efficient synthesis of the surface tantalum hydride ($\equiv SiO)_2Ta^{III}H$ and tris-siloxy tantalum ($\equiv SiO)_3Ta^{III}$ starting from novel tantalum surface species ($\equiv SiO)TaMe_4$ and ($\equiv SiO)_2TaMe_3$. *Organometallics* **2014**, *33*, 1205–1211. [CrossRef]
182. Schrock, R.R.; Meakin, P. Pentamethyl complexes of niobium and tantalum. *J. Am. Chem. Soc.* **1974**, *96*, 5288–5290. [CrossRef]
183. Nilsen, O.; Fjellvåg, H.; Kjekshus, A. Growth of calcium carbonate by the atomic layer chemical vapour deposition technique. *Thin Solid Films* **2004**, *450*, 240–247. [CrossRef]

© 2018 by the authors. Licensee MDPI, Basel, Switzerland. This article is an open access article distributed under the terms and conditions of the Creative Commons Attribution (CC BY) license (http://creativecommons.org/licenses/by/4.0/).

MDPI
St. Alban-Anlage 66
4052 Basel
Switzerland
Tel. +41 61 683 77 34
Fax +41 61 302 89 18
www.mdpi.com

Coatings Editorial Office
E-mail: coatings@mdpi.com
www.mdpi.com/journal/coatings

www.ingramcontent.com/pod-product-compliance
Lightning Source LLC
LaVergne TN
LVHW070602100526
838202LV00012B/542